Logistic Regression Using SAS®

Theory and Application

Second Edition

Paul D. Allison

The correct bibliographic citation for this manual is as follows: Allison, Paul D. 2012. *Logistic Regression Using SAS®: Theory and Application, Second Edition*. Cary, NC: SAS Institute Inc.

Logistic Regression Using SAS®: Theory and Application, Second Edition

ISBN 978-1-60764-995-3 (electronic book)
ISBN 978-1-59994-641-2

SAS Institute Inc., SAS Campus Drive, Cary, North Carolina 27513-2414

1st printing, April 2012

Contents

Preface

It's about time! The first edition of *Logistic Regression Using SAS* had become so outdated that I was embarrassed to see people still using it. In the 13 years since the initial publication, there have been an enormous number of changes and enhancements to the SAS procedures for doing logistic regression and related methods. In fact, I think that the rate of change has accelerated in recent years.

So, as of April 2012, this book is up to date with the latest syntax, features, and options in SAS 9.3. All the output displays use the HTML format, which is now the default. Perhaps the biggest change is that PROC LOGISTIC plays an even larger role than before. That's because it now has a CLASS statement, and it can now estimate the multinomial logit model. Here are some chapter-by-chapter details.

Chapter 2, "Binary Logistic Regression with PROC LOGISTIC: Basics." In the first edition, PROC GENMOD got major attention in this chapter. Now it's barely mentioned. The reason for this change is that PROC LOGISTIC has a CLASS statement, so GENMOD isn't needed until Chapter 8. The CLASS statement in PROC LOGISTIC works a little differently than in other PROCs, so I spend some time explaining the differences. In addition, I now show how to get robust standard errors when estimating the linear probability model with PROC REG. And I demonstrate how to estimate marginal effects for the logistic model with PROC QLIM.

Chapter 3, "Binary Logistic Regression: Details and Options." There are lots of changes and additions in this chapter. As in Chapter 2, GENMOD is out, LOGISTIC is in. For dealing with small samples and separation, I explain two major new features of LOGISTIC: exact logistic regression and penalized likelihood. I show how to use the new ODDSRATIO statement to get interpretable odds ratios when a model contains interactions. And I demonstrate the new EFFECTPLOT statement for getting useful graphs of predicted values, especially when there are non-linearities and interactions. Other additions include: graphs of odds ratios, ROC curves with tests for model differences, a new R-squared measure, new diagnostic graphs, and the use of PROC QLIM to model group comparisons with heteroscedastic errors.

Chapter 4, "Logit Analysis of Contingency Tables." The major change in this chapter is that PROC LOGISTIC is now used to estimate all the contingency table models rather than PROC GENMOD.

Chapter 5, “Multinomial Logit Analysis.” The big change in this chapter is that PROC LOGISTIC is used to estimate the multinomial logit model rather than PROC CATMOD. I also show how to produce predicted values in a new format, and I demonstrate the use of the EFFECTPLOT statement to graph predicted values from the multinomial logit model.

Chapter 6, “Logistic Regression for Ordered Categories.” In this chapter, I show how to use PROC QLIM to model heteroscedasticity in the cumulative logit model.

Chapter 7, “Discrete Choice Analysis.” The first edition used PROC PHREG for all the models. Now PROC LOGISTIC (with the STRATA statement) is used for all conditional logit models except those for ranked data. I also provide an introduction to PROC MDC for more advanced choice models, including the heterogeneous extreme value model, the multinomial probit model, and the nested logit model.

Chapter 8, “Logit Analysis of Longitudinal and Other Clustered Data.” The major addition to this chapter is the use of PROC GLIMMIX to estimate mixed logistic models. Important features of this procedure include the use of Gaussian quadrature to get maximum likelihood estimates and the COVTEST statement, which produces correct *p*-values for testing hypotheses about variance parameters. For PROC GENMOD, I now discuss the use of alternating logistic regression to get an odds ratio parameterization of the correlation structure. PROC LOGISTIC is used instead of PROC PHREG for the fixed effects models. And, finally, I show how to get robust standard errors with PROC SURVEYLOGISTIC.

Chapter 9, “Regression for Count Data.” The negative binomial model is now a standard option in PROC GENMOD rather than a macro. I also discuss zero-inflated models at some length.

Chapter 10, “Loglinear Analysis of Contingency Tables.” Except for new versions of the SAS output, this chapter is virtually unchanged.

Once again, many thanks to my editor, George McDaniel, and to all the other folks at SAS who have been so helpful and patient in getting this new edition into print. In particular, the comments and suggestions of the reviewers—Bob Derr, Mike Patetta, David Schlotzhauer—were crucial in getting things right. Of course, I take full responsibility for any errors that remain.

Chapter 1

Introduction

1.1 What This Book Is About

When I began graduate study at the University of Wisconsin in 1970, categorical data analysis consisted of chi-square tests for cross-tabulated data, a technique introduced around 1900 by the great Karl Pearson. This methodology was viewed with scorn by most of my quantitatively oriented cohorts. It was the province of old fogies who hadn't bothered to learn about REGRESSION ANALYSIS, the new universal tool for social science data analysis. Little did we realize that another revolution was taking place under our noses. By the time I left Wisconsin in 1975, the insanely great new thing was LOGLINEAR ANALYSIS, which made it possible to analyze complicated contingency tables in ways that Karl Pearson never dreamed of. But loglinear analysis was a rather different animal from linear regression and I, for one, never felt entirely comfortable working in the loglinear mode.

In the years after I left Wisconsin, these dissimilar approaches to data analysis came together in the form of LOGIT ANALYSIS, also known as logistic regression analysis. The logit model is essentially a regression model that is tailored to fit a categorical dependent variable. In its most widely used form, the dependent variable is a simple dichotomy, and the independent variables can be either quantitative or categorical. As we shall see, the logit model can be generalized to dependent variables that have more than two categories, both ordered and unordered. Using the method of conditional logit analysis, it can also be extended to handle specialized kinds of data such as discrete-choice applications, matched-pair analysis, and longitudinal data. Logit models for longitudinal data can also be estimated with such methods as generalized estimating equations and logistic regression with random effects.

This book is an introduction to the logit model and its various extensions. Unlike most introductory texts, however, this one is heavily focused on the use of SAS to estimate logit and related models. In my judgment, you can't fully understand and appreciate a new statistical technique without carefully considering the practical details of estimation. To accomplish that, it's necessary to choose a particular software system to carry out the computations. Although there are many good statistical packages for doing logistic regression, SAS is certainly among the best in terms of the range of estimation methods, available features and options, efficiency and stability of the algorithms, and quality of the documentation. I find I can do almost anything I want to do in SAS and, in the process, I encounter few of the annoying software problems that seem to crop up frequently in other packages.

In addition to the logit model, I also write briefly about two alternatives for binary data, the probit model and the complementary log-log model. The last chapter is about loglinear analysis, which is a close cousin to logit analysis. Because some kinds of contingency table analysis are awkward to handle with the logit model, the loglinear model can be a useful alternative. I don't pretend to give a comprehensive treatment of loglinear analysis, however. The emphasis is on how to do it with the GENMOD procedure, and to show examples of the types of applications where a loglinear analysis might be particularly useful. The penultimate chapter is about Poisson regression models for count data. I've included this topic partly for its intrinsic interest, but also because it's a useful preparation for the loglinear chapter. In PROC GENMOD, loglinear analysis is accomplished by way of Poisson regression.

Besides GENMOD, this book includes discussions of the following SAS procedures: LOGISTIC, SURVEYLOGISTIC, GLIMMIX, NLMIXED, QLIM, and MDC. However, this book is not intended to be a comprehensive guide to these SAS procedures. I discuss only those features that are most widely used, most potentially useful, and most likely to cause problems or confusion. You should always consult the official documentation in the *SAS/STAT User's Guide.*

1.2 What This Book Is Not About

This book does *not* cover a variety of categorical data analysis known as Cochran-Mantel-Haenszel (CMH) statistics, for two reasons. First, I have little expertise on the subject and it would be presumptuous for me to try to teach others. Second, while CMH is widely used in the biomedical sciences, there have been few applications in the social sciences. This disuse is not necessarily a good thing. CMH is a flexible and well-founded approach to categorical data analysis—one that I believe has many potential applications. While it accomplishes many of the same objectives as logit analysis, it is best suited to those situations where the focus is on the relationship between one independent variable and one dependent variable, controlling for a limited set of additional variables. Statistical control is accomplished in a nonparametric fashion by stratification on all possible combinations of the control variables. Consequently, CMH is less vulnerable than logistic regression to certain kinds of specification error but at the expense of reduced statistical power. Stokes et al. (2000) give an excellent and thorough introduction to CMH methods as implemented with the FREQ procedure in SAS.

1.3 What You Need to Know

To understand this book, you need to be familiar with multiple linear regression. That means that you should know something about the assumptions of the linear regression model and about estimation of the model via ordinary least squares. Ideally, you should have a substantial amount of practical experience using multiple regression on real data and should feel comfortable interpreting the output from a regression analysis. You should be acquainted with such topics as multicollinearity, residual analysis, variable selection, nonlinearity, interactions, and dummy (indicator) variables. As part of this knowledge, you must certainly know the basic principles of statistical inference: standard errors, confidence intervals, hypothesis tests, *p*-values, bias, efficiency, and so on. In short, a two-semester sequence in statistics ought to provide the necessary statistical foundation for most people.

I have tried to keep the mathematics at a minimal level throughout the book. Except for one section on maximum likelihood estimation (which can be skipped without loss of continuity), there is no calculus and little use of matrix notation. Nevertheless, to simplify the presentation of regression models, I occasionally use the vector notation $\mathbf{\beta x} = \beta_0 + \beta_1 x_1 + \ldots + \beta_k x_k$. While it would be helpful to have some knowledge of maximum likelihood estimation, it's hardly essential. However, you should know the basic properties of logarithms and exponential functions.

With regard to SAS, the more experience you have with SAS/STAT and the SAS DATA step, the easier it will be to follow the presentation of SAS programs. On the other hand, the programs presented in this book are fairly simple and short, so don't be intimidated if you're just beginning to learn SAS.

1.4 Computing

All the computer input and output displayed in this book was produced by and for SAS 9.3. Occasionally, I point out differences between the syntax of SAS 9.3 and earlier releases. I use the following convention for presenting SAS programs: all SAS keywords are in uppercase. All user-specified variable names and data set names are in lowercase. In the main text, both SAS keywords and user-specified variables are in uppercase. In the output displays, nonessential output lines are often edited out to conserve space. If you would like to get a copy of the SAS code and the data sets used in this book, you can download them from my SAS Press author page at http://support.sas.com/publishing/authors/allison.html.

1.5 References

Most of the topics in this book can be found in any one of several textbooks on categorical data analysis. In preparing this book, I have particularly benefited from consulting Agresti (2002), Hosmer and Lemeshow (2000), Long (1997), Fienberg (2007), and Everitt (1992). I have generally refrained from giving references for material that is well established in the textbook literature, but I do provide references for any unusual, nonstandard, or controversial claims. I also give references whenever I think the reader might want to pursue additional information or discussion. This book should not be regarded as a substitute for official SAS documentation. The most complete and up-to-date documentation for the procedures discussed in this book can be found online at http://support.sas.com/documentation.

Chapter 2

Binary Logistic Regression with PROC LOGISTIC: Basics

2.1 Introduction

A great many variables in the social sciences are dichotomous—employed vs. unemployed, married vs. unmarried, guilty vs. not guilty, voted vs. didn't vote. It's hardly surprising, then, that social scientists frequently want to estimate regression models in which the dependent variable is a dichotomy. Nowadays, most researchers are aware that there's something wrong with using ordinary linear regression for a dichotomous dependent variable, and that it's better to use logistic or probit regression. But many of them don't know what it is about linear regression that makes dichotomous variables problematic, and they may have only a vague notion of why other methods are superior.

In this chapter, we focus on logistic regression (a.k.a. logit analysis) as an optimal method for the regression analysis of dichotomous (binary) dependent variables. Along the way, we'll see that logistic regression has many similarities to ordinary linear regression analysis. To understand and appreciate the logistic model, we first need to see why ordinary linear regression runs into problems when the dependent variable is dichotomous.

2.2 Dichotomous Dependent Variables: Example

To make things tangible, let's start with an example. Throughout this chapter, we'll be examining a data set consisting of 147 death penalty cases in the state of New Jersey. In all of these cases, the defendant was convicted of first-degree murder with a recommendation by the prosecutor that a death sentence be imposed. Then a penalty trial was conducted to determine whether the defendant would receive a sentence of death or life imprisonment. Our dependent variable DEATH is coded 1 for a death sentence, and 0 for a life sentence. The aim is to determine how this outcome was influenced by various characteristics of the defendant and the crime.

Many potential independent variables are available in the data set, but let's consider three of special interest:

BLACKD	Coded 1 if the defendant was black, otherwise 0.
WHITVIC	Coded 1 if the victim was white, otherwise 0.
SERIOUS	A rating of the seriousness of the crime, as evaluated by a panel of judges.

The variable SERIOUS was developed in an auxiliary study in which panels of trial judges were given written descriptions of each of the crimes in the original data set. These descriptions did not mention the race of the defendant or the victim. Each judge evaluated 14 or 15 cases and ranked them from least serious to most serious. Each case was ranked by four to six judges. As used in this chapter, the SERIOUS score is the average ranking given to each case, ranging from 1 (least serious) to 15 (most serious).

Using the REG procedure in SAS, I estimated a linear regression model that uses DEATH as the dependent variable and the other three as independent variables. The SAS code is

```
PROC REG DATA=penalty;
  MODEL death=blackd whitvic serious;
RUN;
```

Results are shown in Output 2.1. Neither of the two racial variables have coefficients that are significantly different from 0. Not surprisingly, the coefficient for SERIOUS is highly significant—more serious crimes are more likely to get the death penalty.

Should we trust these results, or should we ignore them because the statistical technique is incorrect? To answer that question we need to see *why* linear regression is regarded as inappropriate when the dependent variable is a dichotomy. That's the task of the next section.

Output 2.1 Linear Regression of Death Penalty on Selected Independent Variables

Number of Observations Read	147
Number of Observations Used	147

Analysis of Variance					
Source	**DF**	**Sum of Squares**	**Mean Square**	**F Value**	**Pr > F**
Model	3	2.61611	0.87204	4.11	0.0079
Error	143	30.37709	0.21243		
Corrected Total	146	32.99320			

Root MSE	0.46090	**R-Square**	0.0793
Dependent Mean	0.34014	**Adj R-Sq**	0.0600
Coeff Var	135.50409		

Parameter Estimates					
Variable	DF	Parameter Estimate	Standard Error	t Value	Pr > |t|
Intercept	1	-0.05492	0.12499	-0.44	0.6610
blackd	1	0.12197	0.08224	1.48	0.1403
whitvic	1	0.05331	0.08411	0.63	0.5272
serious	1	0.03840	0.01200	3.20	0.0017

2.3 Problems with Ordinary Linear Regression

Not so long ago, it was common to see published research that used ordinary least squares (OLS) linear regression to analyze dichotomous dependent variables. Some people didn't know any better. Others knew better, but didn't have access to good software for alternative methods. Now, every major statistical package includes a procedure for logistic regression, so there's no excuse for applying inferior methods. No reputable social science journal would publish an article that used OLS regression with a dichotomous dependent variable.

Should all the earlier literature that violated this prohibition be dismissed? Actually, most applications of OLS regression to dichotomous variables give results that are qualitatively quite similar to results obtained using logistic regression. There are exceptions, of course, so I certainly wouldn't claim that there's no need for logistic regression. But as an approximate method, OLS linear regression does a surprisingly good job with dichotomous variables, despite clear-cut violations of assumptions.

What are the assumptions that underlie OLS regression? While there's no single set of assumptions that justifies linear regression, the list in the box below is fairly standard. To keep things simple, I've included only a single independent variable x, and I've presumed that x is "fixed" across repeated samples (which means that every sample has the same set of x values). The i subscript distinguishes different members of the sample.

Assumptions of the Linear Regression Model

1. $y_i = \alpha + \beta x_i + \varepsilon_i$
2. $\text{E}(\varepsilon_i) = 0$
3. $\text{var}(\varepsilon_i) = \sigma^2$
4. $\text{cov}(\varepsilon_i, \varepsilon_j) = 0$
5. $\varepsilon_i \sim Normal$

Assumption 1 says that y is a linear function of x plus a random disturbance term ε, for all members of the sample. The remaining assumptions all say something about the distribution of ε. What's important about assumption 2 is that $\text{E}(\varepsilon)$ (the expected value of ε) does *not* vary with x, implying that x and ε are uncorrelated. Assumption 3, often called the *homoscedasticity* assumption, says that the variance of ε is the same for all observations. Assumption 4 says that the random disturbance for one observation is uncorrelated with the random disturbance for any other observation. Finally, assumption 5 says that the random disturbance is normally distributed. If all five assumptions are satisfied, ordinary least squares estimates of α and β are unbiased and have minimum sampling variance (minimum variability across repeated samples).

Now suppose that y is a dichotomy with possible values of 1 or 0. It's still reasonable to claim that assumptions 1, 2, and 4 are true. But if 1 and 2 are true for a dichotomy, then 3 and 5 are *necessarily* false. First, let's consider assumption 5. Suppose that y_i=1. Then assumption 1 implies that $\varepsilon_i = 1-\alpha-\beta x_i$. On the other hand, if y_i=0, we have $\varepsilon_i = -\alpha-\beta x_i$. Because ε_i can only take on two values, it's impossible for it to have a normal distribution (which has a continuum of values and no upper or lower bound). So assumption 5 must be rejected.

To evaluate assumption 3, it's helpful to do a little preliminary algebra. The expected value of y_i is, by definition,

$$E(y_i) = 1 \times \Pr(y_i = 1) + 0 \times \Pr(y_i = 0).$$

If we define $p_i = \Pr(y_i=1)$, this reduces to

$$E(y_i) = p_i$$

In general, for any dummy variable, its expected value is just the probability that it is equal to 1. But assumptions 1 and 2 also imply another expression for this expectation. Taking the expected values of both sides of the equation in assumption 1, we get

$$\begin{aligned} E(y_i) &= E(\alpha + \beta x_i + \varepsilon_i) \\ &= E(\alpha) + E(\beta x_i) + E(\varepsilon_i) \\ &= \alpha + \beta x_i . \end{aligned}$$

Putting these two results together, we get

$$p_i = \alpha + \beta x_i, \tag{2.1}$$

which is sometimes called the *linear probability model*. As the name suggests, this model says that the probability that y=1 is a linear function of x. Regression coefficients have a straightforward interpretation under this model. A 1-unit change in x produces a change of β in the probability that y=1. In Output 2.1, the coefficient for SERIOUS was .038. So we can say that each 1-point increase in the SERIOUS scale (which ranges from 1 to 15) is associated with an increase of .038 in the probability of a death sentence, controlling for the other variables in the model. The BLACKD coefficient of .12 tells us that the estimated probability of a death sentence for black defendants is .12 higher than for nonblack defendants, controlling for other variables.

Now let's consider the variance of ε_i. Because x is treated as fixed, the variance of ε_i is the same as the variance of y_i. In general, the variance of a dummy variable is $p_i(1-p_i)$. Therefore, we have

$$\text{var}(\varepsilon_i) = p_i(1-p_i) = (\alpha + \beta x_i)(1 - \alpha - \beta x_i).$$

We see, then, that the variance of ε_i must be different for different observations and, in particular, it varies as a function of x. The disturbance variance is at a maximum when p_i=.5 and gets small when p_i is near 1 or 0.

We've just shown that a dichotomous dependent variable in a linear regression model necessarily violates assumptions of homoscedasticity (assumption 3) and normality (assumption 5) of the error term. What are the consequences? Not as serious as you might think. First of all, we don't need these assumptions to get *unbiased* estimates. If just assumptions 1 and 2 hold, ordinary least squares will produce unbiased estimates of α and β. Second, the normality assumption is not needed if the sample is reasonably large. The central limit theorem assures us that coefficient estimates will have a distribution that is approximately normal even when ε is *not* normally distributed. That means that we can still use a normal table to calculate *p*-values and confidence intervals. If the sample is small, however, these approximations could be poor.

Violation of the homoscedasticity assumption has two undesirable consequences. First, the coefficient estimates are no longer *efficient*. In statistical terminology, this means that there are alternative methods of estimation with smaller standard errors. Second, and more serious, the standard error estimates are no longer *consistent* estimates of the true standard errors. That means that the estimated standard errors could be biased (either upward or downward) to unknown degrees. And because the standard errors are used in calculating test statistics, the test statistics could also be problematic.

Fortunately, the potential problems with standard errors and test statistics are easily fixed. Beginning with SAS 9.2, PROC REG offers a *heteroscedasticity consistent covariance estimator* that uses the method of Huber (1967) and White (1980), sometimes known as the "sandwich" method because of the structure of the matrix formula. This method produces consistent estimates of the standard errors even when the homoscedasticity assumption is violated. To implement the method in PROC REG, simply put the option HCC on the MODEL statement:

```
PROC REG DATA=penalty;
  MODEL death=blackd whitvic serious / HCC;
RUN;
```

Now, in addition to the output in Output 2.1, we get the corrected standard errors, *t*-statistics, and *p*-values shown in Output 2.2. In this case, the correction for heteroscedasticity makes almost no difference in the results.

Output 2.2 Linear Regression of Death Penalty with Correction for Heteroscedasticity

Parameter Estimates								
						Heteroscedasticity Consistent		
Variable	DF	Parameter Estimate	Standard Error	t Value	Pr > |t|	Standard Error	t Value	Pr > |t|
Intercept	1	-0.05492	0.12499	-0.44	0.6610	0.11959	-0.46	0.6468
blackd	1	0.12197	0.08224	1.48	0.1403	0.08197	1.49	0.1390
whitvic	1	0.05331	0.08411	0.63	0.5272	0.08315	0.64	0.5224
serious	1	0.03840	0.01200	3.20	0.0017	0.01140	3.37	0.0010

Although the HCC standard errors are an easy fix, be aware that they have inherently more sampling variability than conventional standard errors (Kauermann and Carroll 2001), and may be especially unreliable in small samples. For large samples, however, they should be quite satisfactory.

In addition to these technical difficulties, there is a more fundamental problem with the assumptions of the linear model. We've seen that when the dependent variable is a dichotomy, assumptions 1 and 2 imply the linear probability model

$$p_i = \alpha + \beta x_i$$

While there's nothing intrinsically wrong with this model, it's a bit implausible, especially if *x* is measured on a continuum. If *x* has no upper or lower bound, then for any value of β there are values of *x* for which p_i is either greater than 1 or less than 0. In fact, when estimating a linear probability model by OLS, it's quite common for predicted values generated by the model to be outside the (0, 1) interval. (That wasn't a problem with the regression in Output 2.1, which implied predicted probabilities ranging from .03 to .65.) Of course, it's impossible for the true values (which are probabilities) to be greater than 1 or less than 0. So the only way the model could be true is if a ceiling and floor are somehow imposed on p_i, leading to considerable awkwardness both theoretically and computationally.

These problems with the linear model led statisticians to develop alternative approaches that make more sense conceptually and also have better statistical properties. The most popular of these approaches is the logistic model, which is estimated by maximum likelihood. Before considering the full model, let's examine one of its key components—the *odds* of an event.

2.4 Odds and Odds Ratios

To appreciate the logistic model, it's helpful to have an understanding of *odds* and *odds ratios.* Most people regard probability as the "natural" way to quantify the chances that an event will occur. We automatically think in terms of numbers ranging from 0 to 1, with a 0 meaning that the event will certainly not occur and a 1 meaning that the event certainly will occur. But there are other ways of representing the chances of event, one of which—the odds—has a nearly equal claim to being "natural."

Widely used by professional gamblers, the odds of an event is the ratio of the expected number of times that an event will occur to the expected number of times it will not occur. An odds of 4 means we expect 4 times as many occurrences as non-occurrences. An odds of 1/5 means that we expect only one-fifth as many occurrences as non-occurrences. In gambling circles, odds are sometimes expressed as, say, "5 to 2," but that corresponds to the single number 5/2.

There is a simple relationship between probabilities and odds. If p is the probability of an event and O is the odds of the event, then

$$O = \frac{p}{1-p} = \frac{\textit{probability of event}}{\textit{probability of no event}}$$

$$p = \frac{O}{1+O}. \qquad (2.2)$$

This functional relationship is illustrated in Table 2.1.

Table 2.1 Relationship between Odds and Probability

Probability	Odds
.1	.11
.2	.25
.3	.43
.4	.67
.5	1.00
.6	1.50
.7	2.33
.8	4.00
.9	9.00

Note that odds less than 1 correspond to probabilities below .5, while odds greater than 1 correspond to probabilities greater than .5. Like probabilities, odds have a lower bound of 0. But unlike probabilities, there is no upper bound on the odds.

Why do we need the odds? Because it's a more sensible scale for multiplicative comparisons. If I have a probability of .30 of voting in the next election, and your probability of voting is .60, it's meaningful to claim that your probability is twice as great as mine. But if my probability is .60, it's impossible for your probability to be twice as great. There's no problem on the odds scale, however. A probability of .60 corresponds to odds of .60/.40=1.5. Doubling that yields odds of 3. Converting back to probabilities gives us 3/(1+3)=.75.

This leads us to the odds ratio, which is a widely used measure of the relationship between two dichotomous variables. Consider Table 2.2, which shows the cross-tabulation of race of defendant by death sentence for the 147 penalty-trial cases. The numbers in the table are the actual numbers of cases that have the stated characteristics.

Table 2.2 Death Sentence by Race of Defendant for 147 Penalty Trials

	Blacks	Non-blacks	Total
Death	28	22	50
Life	45	52	97
Total	73	74	147

Overall, the estimated odds of a death sentence are 50/97= .52. For blacks, the odds are 28/45 = .62. For nonblacks, the odds are 22/52 = .42. The ratio of the black odds to the nonblack odds is 1.47. We may say, then, that the odds of a death sentence for blacks are 47% greater than for nonblacks. Note that the odds ratio in a 2×2 table is also equal to the *cross-product ratio*, which is the product of the two main-diagonal frequencies divided by the product of the two off-diagonal frequencies. In this case, we have $(52 \times 28)/(22 \times 45) = 1.47$.

Of course, we can also say that the odds of a death sentence for nonblacks are 1/1.47 = .63 times the odds of a death sentence for blacks. Similarly, the odds of a *life sentence* for blacks are .63 times the odds for nonblacks. So, depending on which categories we're comparing, we either get an odds ratio greater than 1 or its reciprocal, which is less than 1.

Implicit in much of the contemporary literature on categorical data analysis is the notion that odds ratios (and various functions of them) are less sensitive to changes in the marginal frequencies (for example, the total number of death and life sentences) than other measures of association. In this sense, they are frequently regarded as fundamental descriptions of the relationship between the variables of interest. As we shall see, odds ratios are directly related to the parameters in the logistic regression model.

2.5 The Logistic Regression Model

Now we're ready to introduce the *logistic regression model*, otherwise known as the *logit model*. As we discussed earlier, a major problem with the linear probability model is that probabilities are bounded by 0 and 1, but linear functions are inherently unbounded. The solution is to transform the probability so that it's no longer bounded.

Transforming the probability to an odds removes the upper bound. If we then take the logarithm of the odds, we also remove the lower bound. Setting the result equal to a linear function of the explanatory variables, we get the logistic model. For k explanatory variables and $i = 1,\ldots, n$ individuals, the model is

$$\log\left[\frac{p_i}{1-p_i}\right] = \alpha + \beta_1 x_{i1} + \beta_2 x_{i2} + \ldots + \beta_k x_{ik} \qquad (2.3)$$

where p_i is, as before, the probability that y_i=1. The expression on the left-hand side is usually referred to as the *logit* or *log-odds*. (Natural logarithms are used throughout this book. However, the only consequence of switching to base-10 logarithms would be to change the intercept α.) As in ordinary linear regression, the x's may be either quantitative variables or dummy (indicator) variables.

Unlike the usual linear regression model, there is no random disturbance term in the equation for the logistic model. That doesn't mean that the model is deterministic because there is still room for random variation in the probabilistic relationship between p_i and y_i. Nevertheless, as we shall see later, problems may arise if there is unobserved heterogeneity in the sample.

We can solve the logit equation for p_i to obtain

$$p_i = \frac{\exp(\alpha + \beta_1 x_{i1} + \beta_2 x_{i2} + \ldots + \beta_k x_{ik})}{1 + \exp(\alpha + \beta_1 x_{i1} + \beta_2 x_{i2} + \ldots + \beta_k x_{ik})} \qquad (2.4)$$

Exp(x) is the exponential function, equivalent to e^x. In turn, e is the exponential constant, approximately equal to 2.71828. Its defining property is that log(e^x)=x. We can simplify further by dividing both the numerator and denominator by the numerator itself:

$$p_i = \frac{1}{1+\exp(-\alpha - \beta_1 x_{i1} - \beta_2 x_{i2} - \ldots - \beta_k x_{ik})} \tag{2.5}$$

This equation has the desired property that no matter what values we substitute for the β's and the x's, p_i will always be a number between 0 and 1.

If we have a single x variable with $\alpha = 0$ and $\beta = 1$, the equation can be graphed to produce the S-shaped curve in Figure 2.1. As x gets large or small, p gets close to 1 or 0 but is never equal to these limits. From the graph, we see that the effect of a unit change in x depends on where you start. When p is near .50, the effect is large; but when p is near 0 or 1, the effect is small. More specifically, the slope of the curve is given by the derivative of p_i with respect to the covariate x_i.

$$\frac{\partial p_i}{\partial x_i} = \beta p_i (1 - p_i). \tag{2.6}$$

This is known as the "marginal effect" of x on the event probability, which will be discussed in Section 2.8. When β=1 and p=.5, a 1-unit increase in x produces an increase of .25 in the probability. When β is larger, the slope of the S-shaped curve at p=.5 is steeper. When β is negative, the curve is flipped horizontally so that p is near 1 when x is small and near 0 when x is large. The derivative in equation (2.6) also applies when there is more than one x variable, although then it's a partial derivative.

There are alternative models that have similar S-shaped curves, most notably the probit and complementary log-log models. I'll discuss them briefly in the next chapter. But, for several reasons, the logistic model is more popular:

- The coefficients have a simple interpretation in terms of odds ratios.
- The logistic model is intimately related to the loglinear model, which is discussed in Chapter 10.
- The logistic model has desirable sampling properties, which are discussed in Section 3.13.
- The model can be easily generalized to allow for multiple, unordered categories for the dependent variable.

Figure 2.1 Graph of Logistic Model for a Single Explanatory Variable

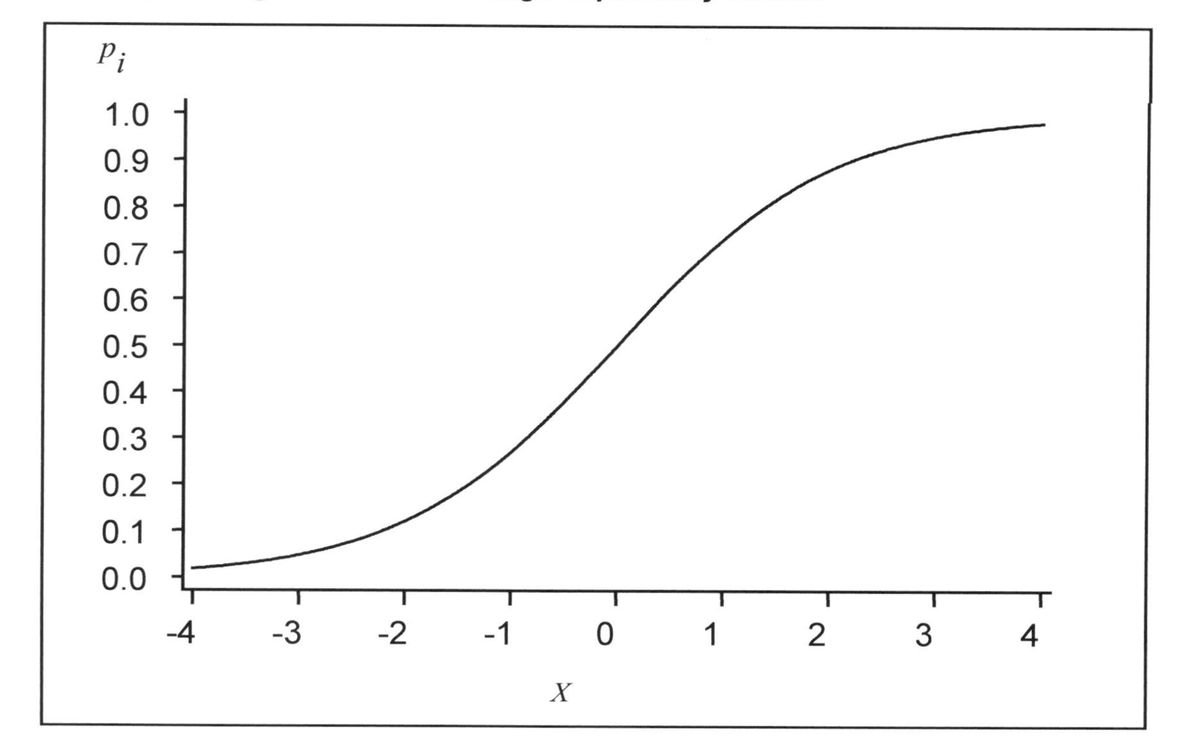

2.6 Estimation of the Logistic Model: General Principles

Now that we have a model for dichotomous dependent variables, the next step is to use sample data to estimate the coefficients. How that's done depends on the type of data you're working with. If you have *grouped data*, there are three readily available methods: ordinary least squares, weighted least squares, and maximum likelihood.

Grouped data occurs when the explanatory variables are all discrete and the data is arrayed in the form of a contingency table. We'll see several examples of grouped data in Chapter 4. Grouped data can also occur when data are collected from naturally occurring groups. For example, suppose that the units of analysis are business firms and the dependent variable is the probability that an employee is a full-time worker. Let P_i be the observed proportion of employees who work full-time in firm *i*. To estimate a logistic model by OLS, we could simply take the logit transformation of *P*, which is log[*P*/(1-*P*)], and regress the result on characteristics of the firm and on the average characteristics of the employees. A weighted least squares (WLS) analysis would be similar, except that the data would be weighted to adjust for heteroscedasticity. The SAS procedure CATMOD does WLS estimation for grouped data (in addition to maximum likelihood).

Maximum likelihood (ML) is the third method for estimating the logistic model for grouped data and the *only* method in general use for *individual-level data*. With individual-level data, we simply observe a dichotomous dependent variable for each individual along with measured characteristics of the individual. OLS and WLS can't be used with this kind of data unless the data can be grouped in some way. If y_i can only have values of 1 and 0, it's impossible to apply the logit transformation—you get either minus infinity or plus infinity. To put it another way, any transformation of a dichotomy is still a dichotomy.

Maximum likelihood is a very general approach to estimation that is widely used for all sorts of statistical models. You may have encountered it before with loglinear models, latent variable models, or event history models. There are two reasons for this popularity. First, ML estimators are known to have good properties in large samples. Under fairly general conditions, ML estimators are consistent, asymptotically efficient, and asymptotically normal. Consistency means that as the sample size gets larger the probability that the estimate is within some small distance of the true value also gets larger. No matter how small the distance or how high the specified probability, there is always a sample size that yields an even higher probability that the estimator is within that distance of the true value. One implication of consistency is that the ML estimator is approximately unbiased in large samples. Asymptotic efficiency means that, in large samples, the estimates will have standard errors that are, approximately, at least as small as those for any other estimation method. And, finally, the sampling distribution of the estimates will be approximately normal in large samples, which means that you can use the normal and chi-square distributions to compute confidence intervals and *p*-values.

All these approximations get better as the sample size gets larger. The fact that these desirable properties have only been proven for large samples does *not* mean that ML has bad properties for small samples. It simply means that we usually don't *know* exactly what the small-sample properties are. And in the absence of attractive alternatives, researchers routinely use ML estimation for both large and small samples. Although I won't argue against that practice, I do urge caution in interpreting *p*-values and confidence intervals when samples are small. Despite the temptation to accept *larger p*-values as evidence against the null hypothesis in small samples, it is actually more reasonable to demand *smaller* values to compensate for the fact that the approximation to the normal or chi-square distributions may be poor.

The other reason for ML's popularity is that it is often straightforward to derive ML estimators when there are no other obvious candidates. One case that ML handles very nicely is data with categorical dependent variables.

The basic principle of ML is to choose as estimates those parameter values that, if true, would maximize the probability of observing what we have, in fact, observed. There are two steps to this: (1) write down an expression for the probability of the data as a function of the unknown parameters, and (2) find the values of the unknown parameters that make the value of this expression as large as possible.

The first step is known as constructing the *likelihood function*. To accomplish this you must specify a model, which amounts to choosing a probability distribution for the dependent variable and choosing a functional form that relates the parameters of this distribution to the values of the explanatory variables. In the case of the logistic model, the dichotomous dependent variable is presumed to have a binomial distribution with a single "trial" and a probability of "success" given by p_i. Then p_i is assumed to depend on the explanatory variables according to equation (2.3), which is the logistic model. Finally, we assume that the observations are independent across individuals.

The second step—maximization—typically requires an iterative numerical method, which means that it involves successive approximations. Such methods are usually more computationally demanding than a non-iterative method like ordinary least squares. For those who are interested, I will work through the basic mathematics of constructing and maximizing the likelihood function in Chapter 3. Here I focus on some of the practical aspects of ML estimation with SAS.

2.7 Maximum Likelihood Estimation with PROC LOGISTIC

The most popular SAS procedure for doing ML estimation of the logistic regression model is PROC LOGISTIC. SAS has several other procedures that will also do this, and we will meet some of them in later chapters. But LOGISTIC has been around the longest, and it has the largest set of features that are likely to be used by most data analysts.

Let's estimate a logistic model analogous to the linear probability model that we examined in Section 2.2. Minimal SAS code for this model is

```
PROC LOGISTIC DATA=penalty;
  MODEL death(EVENT='1')=blackd whitvic serious;
RUN;
```

Of course, there are also numerous options and special features that we'll consider later.

One option that I've specified in the MODEL statement is EVENT='1', after the dependent variable. The default in LOGISTIC is to estimate a model predicting the *lowest* value of the dependent variable. Consequently, if I had omitted EVENT='1', the result would be a logistic model predicting the probability that the dependent variable DEATH is equal to 0. The EVENT='1' option reverses this so that the model predicts the probability that the dependent variable is equal to 1. (The single quotes around 1 are necessary because PROC LOGISTIC treats the dependent variable as a character variable rather than as a numeric variable.)

An equivalent (and popular) way to accomplish this is to use the option DEATH(DESCENDING), which tells LOGISTIC to model the "higher" value of DEATH rather than the lower. But what is considered higher rather than lower can depend on other options that are chosen, so it's safer to be explicit about which value of the dependent variable is to be modeled. If you forget the EVENT='1' option, the only consequence is to change the signs of the coefficients. As long as you realize what you've done, you shouldn't need to rerun the model.

Results are shown in Output 2.3. The "Model Information" table is pretty straightforward, except for the "Optimization Technique," which is reported as Fisher's scoring. This is the numerical method used to maximize the likelihood function, and we will see how it works in Chapter 3. After the "Response Profile" table, we see the message "Probability modeled is death=1." If we had not used the EVENT='1' option, this would have said "death=0", so it's important to check this so that you correctly interpret the signs of the coefficients. Next we are told that the convergence criterion was satisfied, which is a good thing. Iterative methods don't always converge, and we'll talk about why that happens and how to deal with it in Chapter 3.

The next table reports three different "Model Fit Statistics:" AIC, SC, and -2 Log L. Values of these fit statistics are displayed for two different models, a model with an intercept but no covariates (predictors), and a model that includes all the specified predictors (covariates). Usually, we can ignore the INTERCEPT ONLY column. The most fundamental of the fit statistics, -2 Log L, is simply the maximized value of the logarithm of the likelihood function multiplied by –2. We'll see how this number is obtained in Chapter 3. Higher values of -2 Log L mean a worse fit to the data. But keep in mind that the overall magnitude of this statistic is heavily dependent on the number of observations. Furthermore, there is no absolute standard for what's a good fit, so one can only use this statistic to compare different models fit to the same data set.

The problem with -2 Log L is that models with more covariates tend to fit better by chance alone. The other two fit statistics avoid this problem by penalizing models that have more covariates. Akaike's Information Criterion (AIC) is calculated as

$$AIC = -2\log L + 2k$$

where k is the number of parameters (including the intercept). So, in this example, there are four parameters, which adds 8 to the -2 Log L.

The Schwarz Criterion (SC), also known as the Bayesian Information Criterion (BIC), gives a more severe penalization for additional parameters:

$$SC = -2\log L + k\log n$$

where n is the sample size. In this example, $n = 147$ and $\log n = 4.99$. So, to get the SC value, we add 4 times 4.99 = 19.96 to the -2 Log L.

Both of the penalized statistics can be used to compare models with different sets of covariates. The models being compared do not have to be nested in the sense of one model being a special case of another. However, these statistics cannot be used to construct a formal hypothesis test, so the comparison is only informal.

The next table is "Testing Global Null Hypothesis: BETA=0." Within this table there are three chi-square statistics with values of 12.206, 11.656, and 10.8211. All three statistics are testing the same null hypothesis—that *all* the explanatory variables have coefficients of 0. (In a linear regression, this hypothesis is usually tested by means of an overall F-test.) The three degrees of freedom for each statistic correspond to the three coefficients for the independent variables. In this case, the associated p-values are around .01, so we can reject the null hypothesis and conclude that at least one of the coefficients is not 0.

Output 2.3 PROC LOGISTIC Output for Death Penalty Data

Model Information	
Data Set	PENALTY
Response Variable	death
Number of Response Levels	2
Model	binary logit
Optimization Technique	Fisher's scoring

Number of Observations Read	147
Number of Observations Used	147

Response Profile		
Ordered Value	death	Total Frequency
1	1	50
2	0	97

Probability modeled is death=1.

Model Convergence Status
Convergence criterion (GCONV=1E-8) satisfied.

Model Fit Statistics		
Criterion	Intercept Only	Intercept and Covariates
AIC	190.491	184.285
SC	193.481	196.247
-2 Log L	188.491	176.285

Testing Global Null Hypothesis: BETA=0			
Test	Chi-Square	DF	Pr > ChiSq
Likelihood Ratio	12.2060	3	0.0067
Score	11.6560	3	0.0087
Wald	10.8211	3	0.0127

Analysis of Maximum Likelihood Estimates					
Parameter	DF	Estimate	Standard Error	Wald Chi-Square	Pr > ChiSq
Intercept	1	-2.6516	0.6748	15.4424	<.0001
blackd	1	0.5952	0.3939	2.2827	0.1308
whitvic	1	0.2565	0.4002	0.4107	0.5216
serious	1	0.1871	0.0612	9.3342	0.0022

Odds Ratio Estimates			
Effect	Point Estimate	95% Wald Confidence Limits	
blackd	1.813	0.838	3.925
whitvic	1.292	0.590	2.832
serious	1.206	1.069	1.359

Association of Predicted Probabilities and Observed Responses			
Percent Concordant	67.2	Somers' D	0.349
Percent Discordant	32.3	Gamma	0.351
Percent Tied	0.5	Tau-a	0.158
Pairs	4850	c	0.675

Why do we need three chi-square statistics? The first one is the *likelihood ratio chi-square* obtained by comparing the log-likelihood for the fitted model with the log-likelihood for a model with *no* explanatory variables (intercept only). It is calculated by taking twice the positive difference in the two log-likelihoods. In fact, LOGISTIC reports –2 Log L for each of those models, and the chi-square is just the difference between those two numbers. The *score* statistic is a function of the first and second derivatives of the log-likelihood function under the null hypothesis. The Wald statistic is a function of the coefficients and their covariance matrix. In large samples, there's no reason to prefer any one of these statistics, and they will generally be quite close in value. In small samples or samples with extreme data patterns, there is some evidence that the likelihood ratio chi-square is superior (Jennings 1986), especially when compared with the Wald test.

Next we get to the heart of the matter—the "Analysis of Maximum Likelihood Estimates." As with linear regression, we get coefficient estimates, their estimated standard errors, and test-statistics for the null hypotheses that each coefficient is equal to 0. The test statistics are labeled “Wald Chi-Square.” They are calculated by dividing each coefficient by its standard error and squaring the result. If we omitted the squaring operation (as many software packages do), we could call them *z* statistics, and they would have a standard normal distribution under the null hypothesis. In that case, the *p*-values calculated from a normal table would be exactly the same as the chi-square *p*-values reported here.

We see that the variable SERIOUS has a highly significant coefficient, while the coefficient for the variable WHITVIC is clearly not significant. With a *p*-value of .13, BLACKD is approaching conventional significance levels but doesn’t quite make it. Now compare these *p*-values with those in Output 2.1 for an ordinary linear regression analysis. While not identical, they are remarkably similar. In this case at least, logistic regression and OLS regression lead us to exactly the same qualitative conclusions. It’s more difficult to compare coefficient estimates across the two methods, however, and I won’t attempt that in this chapter.

The odds ratios in the next table are obtained by simply exponentiating the coefficients in the first column, that is, calculating $\exp(\beta)$. They are very important in the interpretation of logistic regression coefficients, and we will discuss that interpretation in the next section. The 95 percent confidence intervals are obtained as follows. First, we get 95 percent confidence intervals around the original β coefficients in the usual way. That is, we add and subtract 1.96 standard errors. To get confidence intervals around the odds ratios, we exponentiate those upper and lower confidence limits.

The last section of Output 2.3, labeled “Association of Predicted Probabilities and Observed Responses,” is an attempt to measure the explanatory power of the model. I’ll discuss these statistics in Section 3.7.

2.8 Interpreting Coefficients

When logistic regression first became popular, a major complaint by those who resisted its advance was that the coefficients had no intuitive meaning. Admittedly, they’re not as easy to interpret as coefficients in the linear probability model. For the linear probability model, a coefficient of .25 tells you that the predicted probability of the event increases by .25 for every 1-unit increase in the explanatory variable. By contrast, a logit coefficient of .25 tells you that the log-odds increases by .25 for every 1-unit increase in the explanatory variable. But who knows what a .25 increase in the log-odds means?

The basic problem is that the logistic model assumes a nonlinear relationship between the probability and the explanatory variables, as shown in Figure 2.1. The change in the probability for a 1-unit increase in an independent variable varies according to where you start. Things become much simpler, however, if we think in terms of odds rather than probabilities.

In Output 2.3, we saw the estimated β coefficients and their associated statistics in the "Analysis of Maximum Likelihood Estimates" table. Except for their sign, they *are* hard to interpret. Let's look instead at the numbers in the "Odds Ratio Estimates" table which, in this example, are obtained from the parameter estimates by computing e^{β}. (When there are CLASS variables in the model, the odds ratios may be computed differently, depending on the parameterization). These might be better described as *adjusted* odds ratios because they control for other variables in the model. Recall that BLACKD has a value of 1 for black defendants and 0 for everyone else. The odds ratio of 1.813 tells us that the predicted odds of a death sentence for black defendants are 1.813 times the odds for nonblack defendants. In other words, the odds of a death sentence for black defendants are 81% *higher* than the odds for other defendants. This compares with an *unadjusted* odds ratio of 1.47 found in Table 2.2. Although the adjusted odds ratio for BLACKD is not statistically significant (the 95% confidence interval includes 1, corresponding to no effect), it is still our best estimate of the effect of this variable.

For the dummy variable WHITVIC, which indicates white victim, the odds ratio is 1.292. This implies that the predicted odds of death are about 29% higher when the victim is white compared to the odds when the victim is not white. Of course, the coefficient is far from statistically significant so we wouldn't want to put much confidence in this value. What about the coefficient for the variable SERIOUS, which *is* statistically significant at the .01 level? Recall that this variable is measured on a 15-point scale. For quantitative variables, it's helpful to subtract 1 from the odds ratio and multiply by 100, that is, calculate $100(e^{\beta}-1)$. This tells us the *percent change* in the odds for each 1-unit increase in the independent variable. In this case, we find that a 1-unit increase in the SERIOUS scale is associated with a 21% increase in the predicted odds of a death sentence. Note that if a β coefficient is significantly different from 0, then the corresponding odds ratio is significantly different from 1. There is no need for a separate test for the odds ratio.

Interpretation of coefficients in terms of odds ratios is certainly the easiest way to approach the logistic model. On the other hand, odds ratios can sometimes be misleading if the probabilities are near 1 or 0. Suppose that in a wealthy suburban high school, the probability of graduation is .99, which corresponds to odds of 99. When financial aid is increased for needy students, the probability of graduation goes up to .995, which implies odds of 199. Apparently, the odds of graduation have more than *doubled* under the new program even though only half a percent more students are graduating. Is this a meaningful increase? That depends on many nonstatistical issues.

For those who insist on interpreting logistic models in terms of probabilities, there are several graphical and tabular methods available (Long 1997). Perhaps the simplest approach is to make use of equation (2.6):

$$\frac{\partial p_i}{\partial x_i} = \beta p_i(1 - p_i).$$

This equation says that the change in the probability for a 1-unit increase in x depends on the logistic regression coefficient for x, as well as on the value of the probability itself. For this to be practically useful, we need to know what probability we are starting from. If we have to choose one value, the most natural is the overall proportion of cases that have the event. In our example, 50 out of 147 defendants got the death penalty, so the overall proportion is .34. Taking .34 times 1–.34, we get .224. We can multiply each of the coefficients in Output 2.3 by .224, and we get:

BLACKD	.133
WHITVIC	.057
SERIOUS	.046

We can then say that, on average, the probability of a death sentence is .133 higher if the defendant is black compared with nonblacks, .057 higher if the victim is white compared with non-white, and .046 higher for a 1-unit increase on the SERIOUS scale. These are sometimes called “marginal effects,” and they are of considerable interest in some fields. Of course, these numbers only give a rough indication of what actually happens for a given change in the x variable. Note, however, that they are very similar to the coefficients obtained with the OLS regression in Output 2.1.

Instead of choosing a single value for p_i, we can calculate a predicted probability for each individual, using equation (2.5). Then we can use the derivative formula to generate marginal effects for each individual. You can easily get them with PROC QLIM (part of the

SAS/ETS product) using the OUTPUT statement with the MARGINAL option. Here's the code:

```
PROC QLIM DATA=penalty;
  ENDOGENOUS death~DISCRETE(DIST=LOGISTIC);
  MODEL death = blackd whitvic serious;
  OUTPUT OUT=a MARGINAL;
PROC PRINT DATA=a(OBS=10);
  VAR meff_p2_blackd meff_p2_whitvic meff_p2_serious;
RUN;
```

QLIM reports coefficients, standard errors, and test statistics (not shown) that are identical to those produced by PROC LOGISTIC. PROC PRINT produces the table shown in Output 2.4, for the first 10 observations in the output data set. For each variable, we get the predicted change in the probability of the death penalty for a 1-unit increase in that variable, for a particular individual based on that individual's predicted probability.

Output 2.4 Marginal Effects Produced by PROC QLIM for the First 10 Cases

Obs	Meff_P2_blackd	Meff_P2_whitvic	Meff_P2_serious
1	0.13068	0.056312	0.041070
2	0.14660	0.063172	0.046073
3	0.08712	0.037541	0.027380
4	0.12615	0.054358	0.039645
5	0.14692	0.063309	0.046173
6	0.10523	0.045347	0.033072
7	0.09613	0.041422	0.030210
8	0.14880	0.064118	0.046763
9	0.12982	0.055942	0.040800
10	0.13682	0.058957	0.042999

2.9 CLASS Variables

As with several other SAS regression procedures, PROC LOGISTIC has a CLASS statement that allows you to specify that a variable should be treated as categorical (nominal). When a CLASS variable is included as an explanatory variable in the MODEL statement, LOGISTIC automatically creates a set of "design variables" to represent the levels of the CLASS variable. Keep in mind that when a predictor variable is an indicator (dummy) variable, like BLACKD or WHITVIC which only have values of 0 or 1, there is no need to declare it to be a CLASS variable. In fact, putting an indicator variable on the CLASS statement can produce misleading results. That's because the CLASS statement might recode the variable in unexpected ways, as we will see. So the CLASS statement should be reserved for categorical variables with more than two categories, or for dichotomous variables that have character values like "yes" and "no."

Here's an example with the death-penalty data. The data set contains the variable CULP, which has the integer values 1 to 5 (5 denotes high culpability and 1 denotes low culpability, based on a large number of aggravating and mitigating circumstances defined by statute). Although we could treat this variable as an interval scale, we might prefer to treat it as a set of categories. To do this, we run the following program:

```
PROC LOGISTIC DATA=penalty;
  CLASS culp /PARAM=REF;
  MODEL death(EVENT='1') = blackd whitvic culp ;
RUN;
```

The CLASS statement declares CULP to be a classification variable. You can have more than one variable on the CLASS statement. The PARAM=REF option tells LOGISTIC to create a set of four dummy (indicator) variables, one for each value of CULP except the highest one (CULP=5). Results are shown in Output 2.5.

Output 2.5 Use of a CLASS Variable in PROC LOGISTIC

Type 3 Analysis of Effects			
Effect	DF	Wald Chi-Square	Pr > ChiSq
blackd	1	7.9141	0.0049
whitvic	1	2.1687	0.1408
culp	4	44.0067	<.0001

Analysis of Maximum Likelihood Estimates						
Parameter		**DF**	**Estimate**	**Standard Error**	**Wald Chi-Square**	**Pr > ChiSq**
Intercept		1	0.5533	0.7031	0.6193	0.4313
blackd		1	1.7246	0.6131	7.9141	0.0049
whitvic		1	0.8385	0.5694	2.1687	0.1408
culp	**1**	1	-4.8670	0.8251	34.7926	<.0001
culp	**2**	1	-3.0547	0.7754	15.5185	<.0001
culp	**3**	1	-1.5294	0.8400	3.3153	0.0686
culp	**4**	1	-0.3610	0.8857	0.1662	0.6835

Because the variable CULP has five possible values, LOGISTIC has created four dummy variables, one for each of the values 1 through 4. As in other procedures that have CLASS variables, the default is to take the highest value as the omitted category. We'll see how to change that in a moment. Thus, each of the four coefficients for CULP is a comparison between that particular value and the highest value. More specifically, each coefficient can be interpreted as the log-odds of the death penalty for that particular value of CULP minus the log-odds for CULP=5, controlling for other variables in the model.

When you have a CLASS variable in a model, LOGISTIC provides an additional table, labeled "Type 3 Analysis of Effects." For variables that are *not* CLASS variables, this table is completely redundant with the standard table below it—the chi-squares and *p*-values are exactly the same. For CLASS variables, on the other hand, it gives us something very useful: a test of the null hypothesis that *all* of the coefficients pertaining to this variable are 0. In other words, it gives us a test of whether CULP has any impact on the probability of the death penalty. In this case, we clearly have strong evidence that CULP makes a difference. What's particularly attractive about this test is that it is invariant to the choice of the omitted category, or even to the choice among very different methods for constructing design variables (which I'll discuss in a moment).

The pattern for the four coefficients is just what one might expect. Defendants with CULP=1 are much less likely to get the death sentence than those with CULP=5. Each increase of CULP is associated with an increase in the probability of a death sentence. Notice also that when CULP is included in the model, the coefficient for BLACKD (black defendant) is much larger than it was in Output 2.3 and is now statistically significant.

If you don't like the default reference category (5 in this example), how do you change it? If you want it to be the lowest value of CULP rather than the highest value, just use

```
CLASS culp / PARAM=REF DESCENDING;
```

If you want it to be some particular value, say 3, use

```
CLASS culp(REF='3') / PARAM=REF;
```

Suppose you want to compare two categories of a CLASS variable, such as CULP=2 and CULP=3. You could always accomplish this by rerunning the model, and making one of those values the reference category. But it's usually easier to use a TEST statement or a CONTRAST statement, and you can have as many of those as you want. The following two statements test the same null hypothesis, that there is no difference in the coefficients for CULP=2 and CULP=3:

```
CONTRAST 'Culp2 vs. Culp3' culp 0 1 -1 0;
TEST culp2=culp3;
```

The CONTRAST statement requires a label, which can be any text enclosed by quotes. This is annoying if you only have a single CONTRAST, but very useful if you have more than one because the label helps you distinguish the different tests in the output. Here's how the CONTRAST statement works: we know that CULP has four coefficients. The code "CULP 0 1 -1 0" tells SAS to multiply the first coefficient by 0, the second coefficient by 1, the third coefficient by -1, and the fourth coefficient by 0, add up the results, and test whether the sum is equal to 0. Of course, this is equivalent to testing whether there is a difference between the second and third coefficients.

The TEST statement is a little more straightforward. To refer to a coefficient, you just append the value of the variable to the variable name itself. A TEST statement may optionally have a label, but the label comes before TEST and cannot have any spaces. Thus, we could have written

```
Culp2_vs_Culp3: TEST culp2=culp3;
```

The output from the CONTRAST and TEST statements is nearly identical (see Output 2.6). In this case, the difference in the log-odds of a death penalty for those in category 2 vs. those in category 3 is not quite significant at the .05 level. Note that this difference (and its statistical significance) is invariant to the choice of the reference category.

Output 2.6 Results from TEST and CONTRAST Statements

Contrast Test Results			
Contrast	DF	Wald Chi-Square	Pr > ChiSq
Culp2 vs. Culp3	1	3.8152	0.0508

Linear Hypotheses Testing Results			
Label	Wald Chi-Square	DF	Pr > ChiSq
Test 1	3.8152	1	0.0508

What happens if you leave off the PARAM=REF option? Unfortunately, the default for the CLASS statement in PROC LOGISTIC is quite different from the default in many other regression procedures, such as GLM, GENMOD, PHREG, or LIFEREG. This could lead some analysts to misinterpret the results. In those procedures, the default is to produce a set of dummy variables, much like we did using the PARAM=REF option. The default in LOGISTIC, on the other hand, is to produce design variables that are sometimes described as analysis of variance coding or effect coding.

Output 2.7 shows what you get when you simply write CLASS CULP without using the PARAM option. The "Class Level Information" table tells us how the four design variables are constructed. Clearly these are not indicator variables, because each one can takes on values of 1, 0, or -1. The first design variable has a value of 1 when CULP=1, a value of -1 when CULP=5, and values of 0 for the other values of CULP. The other three are constructed in a similar fashion, except that the CULP value that is assigned a 1 changes for each design variable.

Output 2.7 Results from CLASS Statement Using Default Design Variables

Class Level Information					
Class	Value	Design Variables			
culp	1	1	0	0	0
	2	0	1	0	0
	3	0	0	1	0
	4	0	0	0	1
	5	-1	-1	-1	-1

Type 3 Analysis of Effects			
Effect	DF	Wald Chi-Square	Pr > ChiSq
blackd	1	7.9141	0.0049
whitvic	1	2.1687	0.1408
culp	4	44.0067	<.0001

Analysis of Maximum Likelihood Estimates						
Parameter		DF	Estimate	Standard Error	Wald Chi-Square	Pr > ChiSq
Intercept		1	-1.4092	0.6066	5.3972	0.0202
blackd		1	1.7246	0.6131	7.9141	0.0049
whitvic		1	0.8385	0.5694	2.1687	0.1408
culp	1	1	-2.9046	0.5007	33.6535	<.0001
culp	2	1	-1.0923	0.4645	5.5304	0.0187
culp	3	1	0.4330	0.5292	0.6696	0.4132
culp	4	1	1.6014	0.6026	7.0625	0.0079

Odds Ratio Estimates			
Effect	Point Estimate	95% Wald Confidence Limits	
blackd	5.610	1.687	18.657
whitvic	2.313	0.758	7.061
culp 1 vs 5	0.008	0.002	0.039
culp 2 vs 5	0.047	0.010	0.215
culp 3 vs 5	0.217	0.042	1.124
culp 4 vs 5	0.697	0.123	3.954

As can be seen in the "Type 3 Analysis of Effects" table in Output 2.7, this alternative coding has no effect on the overall test for CULP. And skipping down to the "Odds Ratio Effects" table, it also has no effect here either. Just as when we use PARAM=REF, each odds ratio compares a particular category with category 5. But there are certainly major changes in the CULP coefficients reported in the "Analysis of Maximum Likelihood Estimates" table, along with their standard errors and *p*-values. What's important to understand is that these coefficients are *not* comparisons between each of the first four values of CULP and value 5. Rather they are comparisons between each category of CULP and the overall average (roughly speaking) of the log-odds of getting the death penalty, adjusting for other variables in the model. Thus, if you are in category 1 of CULP, your log-odds of getting the death penalty is 2.9046 *below* average. If you are in category 4, your log-odds is 1.6014 *above* average.

What about category 5? How does it differ from the average? To get that value, you must add together the coefficients for values 1 through 4 and then change the sign. That ensures that all five coefficients sum to zero. You could do that by hand calculation, but it's a lot easier to let SAS do it with the ESTIMATE statement:

```
ESTIMATE 'coeff for 5' culp -1 -1 -1 -1;
```

This says to take the four coefficients for CULP, multiply each by -1, and then add them together. Results are shown in Output 2.8. We see that those with value 5 of CULP have a death penalty log-odds that is 1.9624 higher than average. We also get a standard error and a *z*-test of the null hypothesis that this effect is 0.

Output 2.8 Results from ESTIMATE Statement

Estimate				
Label	Estimate	Standard Error	z Value	Pr > |z|
coeff for 5	1.9624	0.5264	3.73	0.0002

Some people like this approach to creating design variables, but I'm not a fan. I prefer comparisons with an explicit reference category. PROC LOGISTIC also offers several other methods for creating design variables, including ordinal, polynomial, and orthogonalized versions of all the methods.

2.10 Multiplicative Terms in the MODEL Statement

Regression analysts often want to build models that have *interactions* in which the effect of one variable depends on the level of another variable. The most popular way of doing this is to include a new explanatory variable in the model, one that is the product of the two original variables. With PROC LOGISTIC, rather than creating a new variable in a DATA step, you can specify the product directly in the MODEL statement. For example, some criminologists have argued that black defendants who kill white victims may be especially likely to receive a death sentence. We can test that hypothesis for the New Jersey data with this program:

```
PROC LOGISTIC DATA=penalty;
  MODEL death(EVENT='1') = blackd whitvic culp blackd*whitvic;
RUN;
```

This produces the table in Output 2.9.

Output 2.9 Multiplicative Variables in PROC LOGISTIC

Analysis of Maximum Likelihood Estimates					
Parameter	DF	Estimate	Standard Error	Wald Chi-Square	Pr > ChiSq
Intercept	1	-5.4042	1.1626	21.6066	<.0001
blackd	1	1.8720	1.0463	3.2013	0.0736
whitvic	1	1.0725	0.9877	1.1790	0.2776
culp	1	1.2703	0.1967	41.6883	<.0001
blackd*whitvic	1	-0.3272	1.1781	0.0771	0.7812

With a *p*-value of .78, the product term is clearly not significant and can be excluded from the model. The MODEL statement also has a short-hand notation that allows you specify both the interaction and the two main effects:

```
MODEL death(EVENT='1') = culp blackd|whitvic;
```

The product syntax in LOGISTIC also makes it easy to construct polynomial functions. For example, to estimate a *cubic* equation you can specify a model of the form

```
MODEL y = x x*x x*x*x;
```

Or, equivalently

```
MODEL y = x|x|x;
```

This fits the model $\log(p/(1-p)) = \alpha + \beta_1 x + \beta_2 x^2 + \beta_3 x^3$.

Chapter 3

Binary Logistic Regression: Details and Options

3.1 Introduction

In this chapter, we continue with the binary logit model and its implementation in PROC LOGISTIC. We'll examine several optional features of this procedure, consider some potential problems that might arise, and pursue the details of the estimation process in greater depth. You might get by without this chapter, but you'd miss some useful and enlightening stuff.

3.2 Confidence Intervals

It's standard practice in social science journals to report only *point estimates* and hypothesis tests for the coefficients. Most statisticians, on the other hand, hold that *confidence intervals* give a better picture of the sampling variability of the estimates. PROC LOGISTIC automatically produces 95% confidence intervals for the odds ratios, but you may also want them for the regression coefficients.

Conventional confidence intervals for the coefficients are easily computed by hand. For an approximate 95% confidence interval around a coefficient, you simply add and subtract the standard error multiplied by 1.96. But you can save yourself the trouble by asking LOGISTIC to do it for you. In LOGISTIC, the option in the MODEL statement for conventional (Wald) confidence intervals is CLPARM=WALD. To change that to a 90% interval, put the option ALPHA=.10 on the MODEL statement.

LOGISTIC has another method, called *profile likelihood confidence intervals*, that may produce better approximations especially in smaller samples. This method involves an iterative evaluation of the likelihood function. In LOGISTIC, the model option is CLPARM=PL (for profile likelihood). The profile likelihood method is computationally intensive so you may need to use it sparingly for large samples.

Here's an example of both kinds of confidence intervals using LOGISTIC with the death penalty data set:

```
PROC LOGISTIC DATA=penalty;
  MODEL death(EVENT='1') = blackd whitvic culp / CLPARM=BOTH;
RUN;
```

Besides the usual results, we get the numbers shown in Output 3.1. The intervals produced by the two methods are very similar but not identical.

Output 3.1 Confidence Intervals Produced by PROC LOGISTIC

Parameter Estimates and Profile-Likelihood Confidence Intervals			
Parameter	**Estimate**	**95% Confidence Limits**	
Intercept	-5.2179	-7.2485	-3.5710
blackd	1.6358	0.5289	2.8890
whitvic	0.8476	-0.2190	1.9811
culp	1.2708	0.9187	1.6953

Parameter Estimates and Wald Confidence Intervals			
Parameter	**Estimate**	**95% Confidence Limits**	
Intercept	-5.2179	-7.0404	-3.3954
blackd	1.6358	0.4688	2.8029
whitvic	0.8476	-0.2428	1.9380
culp	1.2708	0.8862	1.6554

You can also get profile likelihood confidence intervals for the odds ratios by putting the CLODDS=PL option on the MODEL statement. Output 3.2 shows the profile likelihood confidence intervals for the model just estimated. The UNIT column indicates how much each independent variable is incremented to produce the estimated odds ratio. The default is 1 unit. For the variable CULP, each 1-point increase on the culpability scale multiplies the odds of a death sentence by 3.564. In SAS 9.3 and later, whenever you use the CLODDS option you also get a graph of the odds ratios by default.

Output 3.2 Odds Ratio Confidence Intervals in PROC LOGISTIC

Odds Ratio Estimates and Profile-Likelihood Confidence Intervals				
Effect	**Unit**	**Estimate**	**95% Confidence Limits**	
blackd	1.0000	5.134	1.697	17.976
whitvic	1.0000	2.334	0.803	7.251
culp	1.0000	3.564	2.506	5.448

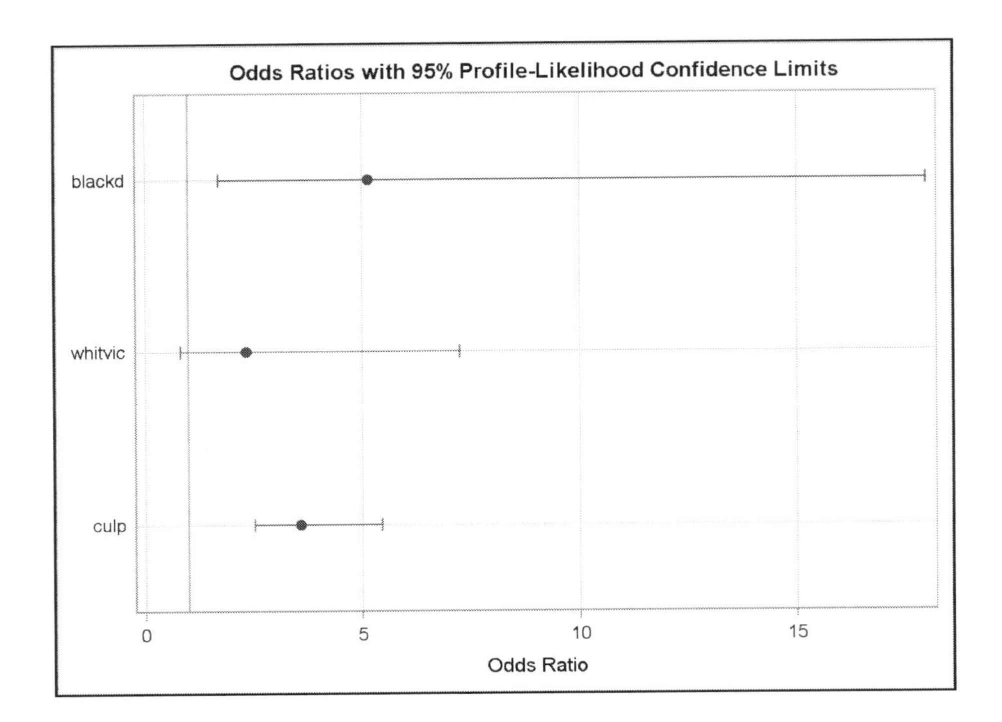

If you want odds ratios for different increments, you could certainly calculate them by hand. If O is the odds ratio for a 1-unit increment, O^k is the odds ratio for a k-unit increment. But it's easier to use the UNITS statement to produce "customized" odds ratios. For example, to get the odds ratio for a 2-unit increase in CULP, include the following statement in the LOGISTIC procedure:

```
UNITS culp=2 / DEFAULT=1;
```

The DEFAULT option tells SAS to print odds ratios and their confidence intervals for a 1-unit increase in each of the other variables in the model.

3.3 Details of Maximum Likelihood Estimation

In Section 2.6, I discussed some general properties of maximum likelihood estimators. Here, we consider the details of the construction and maximization of the likelihood function. Although this section can be skipped without loss of continuity, I strongly encourage you to work through it to the best of your ability. A basic understanding of maximum likelihood for the logistic model can help to remove much of the mystery of the technique. It can also help you understand how and why things sometimes go wrong.

Let's start with some notation and assumptions. We have data for n individuals (i=1,….,n) who are assumed to be statistically independent. For each individual i, the data consists of y_i and $\mathbf{x}_i$, where y_i is a random variable with possible values of 0 and 1, and $\mathbf{x}_i=[1\ x_{i1}\ \ldots\ x_{ik}]'$ is a vector of explanatory variables (the 1 is for the intercept). Vector notation is helpful here; otherwise, the equations get messy. For simplicity, we treat $\mathbf{x}_i$ as a set of fixed constants for each individual rather than as random variables. (We could get equivalent results with $\mathbf{x}_i$ treated as random and y_i expressed conditional on the values of $\mathbf{x}_i$, but that would just complicate the notation.)

Letting p_i be the probability that y_i=1, we assume that the data is generated by a logistic model, which says that

$$p_i = \frac{1}{1+e^{-\boldsymbol{\beta}\mathbf{x}_i}}\ , \tag{3.1}$$

which is equivalent to equation (2.5).

We now construct the likelihood function, which expresses the probability of observing the data in hand as a function of the unknown parameters. The likelihood of observing the values of y for all the observations can be written as

$$L = \Pr(y_1, y_2, \ldots, y_n).$$

Because we are assuming that observations are independent, the overall probability of observing all the y_is can be factored into the product of the individual probabilities:

$$L = \Pr(y_1)\Pr(y_2)\ldots\Pr(y_n) = \prod_{i=1}^{n}\Pr(y_i), \tag{3.2}$$

where Π indicates repeated multiplication.

By definition, $\Pr(y_i=1) = p_i$ and $\Pr(y_i=0) = 1-p_i$. That implies that we can write

$$\Pr(y_i) = p_i^{y_i}(1-p_i)^{1-y_i}. \tag{3.3}$$

In this equation, y_i acts like a switch, turning parts of the formula on or off. When y_i=1, p_i raised to the y_i power is just p_i. But $1-y_i$ is then 0, and $(1-p_i)$ raised to the 0 power is 1. Things are reversed when y_i=0. Substituting equation (3.3) into equation (3.2) and doing a little algebra yields

$$L = \prod_{i=1}^{n} p_i^{y_i}(1-p_i)^{1-y_i} = \prod_{i=1}^{n}\left(\frac{p_i}{1-p_i}\right)^{y_i}(1-p_i)\ .$$

At this point we take the logarithm of both sides of the equation to get

$$\log L = \sum_i y_i \log\left(\frac{p_i}{1-p_i}\right) + \sum_i \log(1-p_i)\,. \tag{3.4}$$

It is generally easier to work with the logarithm of the likelihood function because products get converted into sums and exponents become coefficients. Because the logarithm is an increasing function, whatever maximizes the logarithm of a function will also maximize the original function.

Substituting our expression for the logit model (equation 3.1) into equation (3.4), we get

$$\log L = \sum_i \boldsymbol{\beta}\mathbf{x}_i y_i - \sum_i \log(1+e^{\boldsymbol{\beta}\mathbf{x}_i})\,, \tag{3.5}$$

which is about as far as we can go in simplifying the likelihood function.

This brings us to step 2, choosing values of $\boldsymbol{\beta}$ that make the log-likelihood in equation (3.5) as large as possible. There are many different methods for maximizing functions like this. One well-known approach is to find the derivative of the function with respect to $\boldsymbol{\beta}$, set the derivative equal to 0, and then solve for $\boldsymbol{\beta}$. Taking the derivative of equation (3.5) and setting it equal to 0 gives us:

$$\begin{aligned}\frac{\partial \log L}{\partial \boldsymbol{\beta}} &= \sum_i \mathbf{x}_i y_i - \sum_i \mathbf{x}_i (1+e^{-\boldsymbol{\beta}\mathbf{x}_i})^{-1} \\ &= \sum_i \mathbf{x}_i y_i - \sum_i \mathbf{x}_i \hat{y}_i = 0\end{aligned} \tag{3.6}$$

where

$$\hat{y}_i = \frac{1}{1+e^{-\boldsymbol{\beta}\mathbf{x}_i}}\,,$$

the predicted probability of y for a given value of $\mathbf{x}_i$. Because $\mathbf{x}_i$ is a vector, equation (3.6) is actually a system of k+1 equations, one for each element of $\boldsymbol{\beta}$.

Those familiar with OLS theory may recognize the second line of equation (3.6) as identical to the *normal equations* for the linear model. The difference is that $\hat{y}$ is a linear function of $\boldsymbol{\beta}$ in the linear model but a nonlinear function of $\boldsymbol{\beta}$ in the logit model. Consequently, except in special cases like a single dichotomous x variable, there is no explicit solution to equation (3.6). Instead, we must rely on iterative methods, which amount to successive approximations to the solution until the approximations "converge" to the correct value. Again, there are many different methods for doing this. All produce the same solution, but they differ in such factors as speed of convergence, sensitivity to starting values, and computational difficulty at each iteration.

One of the most widely used iterative methods is the Newton-Raphson algorithm, which can be described as follows: Let $\mathbf{U}(\boldsymbol{\beta})$ be the vector of first derivatives of log *L* with respect to $\boldsymbol{\beta}$ and let $\mathbf{I}(\boldsymbol{\beta})$ be the matrix of second derivatives of log *L* with respect to $\boldsymbol{\beta}$. That is,

$$\mathbf{U}(\boldsymbol{\beta}) = \frac{\partial \log L}{\partial \boldsymbol{\beta}} = \sum_i \mathbf{x}_i y_i - \sum_i \mathbf{x}_i \hat{y}_i$$

$$\mathbf{I}(\boldsymbol{\beta}) = \frac{\partial^2 \log L}{\partial \boldsymbol{\beta} \partial \boldsymbol{\beta}'} = -\sum_i \mathbf{x}_i \mathbf{x}_i' \, \hat{y}_i (1 - \hat{y}_i)$$

The vector of first derivatives $\mathbf{U}(\boldsymbol{\beta})$ is sometimes called the *gradient* or *score*. The matrix of second derivatives $\mathbf{I}(\boldsymbol{\beta})$ is called the *Hessian*. The Newton-Raphson algorithm is then

$$\boldsymbol{\beta}_{j+1} = \boldsymbol{\beta}_j - \mathbf{I}^{-1}(\boldsymbol{\beta}_j)\mathbf{U}(\boldsymbol{\beta}_j) \tag{3.7}$$

where $\mathbf{I}^{-1}$ is the inverse of **I**. To operationalize this algorithm, we need a set of starting values $\boldsymbol{\beta}_0$. PROC LOGISTIC starts by setting all the slope coefficients equal to 0. The intercept is set equal to log[*p*/(1-*p*)] where *p* is the overall proportion of events. These starting values are substituted into the right-hand side of equation (3.7), which yields the result for the first iteration, $\boldsymbol{\beta}_1$. These values are then substituted back into the right-hand side, the first and second derivatives are recomputed, and the result is $\boldsymbol{\beta}_2$. This process is repeated until you get "convergence," which means that what you plug in on the right-hand side is what you get out on the left. In practice, you never get *exactly* the same thing, so it's necessary to have a convergence criterion to judge how close is close enough.

The default criterion in LOGISTIC is somewhat complicated and perhaps not worth our attention. But since every successful run of PROC LOGISTIC reports "Convergence criterion (GCONV=1E-8) satisfied," we should probably consider what that means. Recall that the maximum of the likelihood function is obtained when the first derivative of the log-likelihood, $\mathbf{U}(\boldsymbol{\beta})$, is equal to 0. To test whether that has actually been accomplished, LOGISTIC computes the following quantity at each iteration *j*:

$$\frac{\mathbf{U}(\boldsymbol{\beta}_j)'\mathbf{I}^{-1}(\boldsymbol{\beta}_j)\mathbf{U}(\boldsymbol{\beta}_j)}{|\log L(\boldsymbol{\beta}_j)|+.000001}$$

If that number is less than .00000001, convergence is declared and the algorithm stops.

After the solution $\hat{\boldsymbol{\beta}}$ is found, a byproduct of the Newton-Raphson algorithm is an estimate of the covariance matrix of the coefficients, which is just $-\mathbf{I}^{-1}(\hat{\boldsymbol{\beta}})$. This matrix, which can be printed by putting COVB as an option in the MODEL statement, is often useful for constructing hypothesis tests about linear combinations of coefficients. Estimates of the standard errors of the coefficients are obtained by taking the square roots of the main diagonal elements of this matrix.

3.4 Convergence Problems

As explained in the previous section, maximum likelihood estimation of the logistic model is an iterative process of successive approximations. Usually the process goes smoothly with no special attention needed. Only rarely does it take more than 10 iterations to achieve convergence. However, sometimes the iterative process breaks down so that convergence is not achieved. Dealing with convergence failures can be one of the more frustrating problems encountered by users of logistic regression.

LOGISTIC has a default limit of 25 iterations. If the algorithm hasn't converged by this limit, LOGISTIC issues a warning message and prints out the results at the last iteration. While it's possible to raise the iteration limit (with the MAXITER= option in the MODEL statement), this rarely solves the problem. Models that haven't converged in 25 iterations will usually never converge.

In most cases of convergence failure, the maximum likelihood estimates simply do not exist. Consider the following SAS program, which inputs six observations on *x* and *y* and then performs a logistic regression of *y* on *x*:

```
DATA compsep;
  INPUT x y;
  DATALINES;
1 0
2 0
3 0
4 1
5 1
```

```
6 1
;
PROC LOGISTIC;
  MODEL y(EVENT='1') = x;
RUN;
```

Output 3.3 PROC LOGISTIC Results for Data with Complete Separation

Model Convergence Status
Complete separation of data points detected.

Warning: The maximum likelihood estimate does not exist.
Warning: The LOGISTIC procedure continues in spite of the above warning. Results shown are based on the last maximum likelihood iteration. Validity of the model fit is questionable.

Model Fit Statistics		
Criterion	Intercept Only	Intercept and Covariates
AIC	10.318	4.007
SC	10.110	3.591
-2 Log L	8.318	0.007

Testing Global Null Hypothesis: BETA=0			
Test	Chi-Square	DF	Pr > ChiSq
Likelihood Ratio	8.3105	1	0.0039
Score	4.6286	1	0.0314
Wald	0.1437	1	0.7046

Analysis of Maximum Likelihood Estimates					
Parameter	DF	Estimate	Standard Error	Wald Chi-Square	Pr > ChiSq
Intercept	1	-44.2060	117.8	0.1409	0.7074
x	1	12.6303	33.3127	0.1437	0.7046

In Output 3.3, LOGISTIC reports "Complete separation of data points detected," and then warns us that "the maximum likelihood estimate does not exist." These warning messages also appear in the LOG window. Complete separation means that there is some linear combination of the explanatory variables that perfectly predicts the dependent variable. In this example, whenever x is greater than 3.5, y is equal to 1. Whenever x is less than 3.5, y is equal to 0. So we can perfectly predict y from x. If you examine the iteration history for these data (which you can do by putting the ITPRINT option in the MODEL statement), you'll find that the coefficient for x gets larger at every iteration. After eight iterations, LOGISTIC detects complete separation and calls a halt to the iterations. The results at this iteration are reported in the "Analysis of Maximum Likelihood Estimates" table. There we see that the coefficient for x is 12.63, which is an extremely large value. The estimated standard error is much larger, leading to a tiny chi-square and a high p-value.

In the "Model Fit Statistics" table, we see that -2 log L for this model (Intercept and Covariates) is .007, very close to 0 which is its lower bound. Figure 3.1 is a graph of the log-likelihood function for coefficient values between 0 and 10. We see that the log-likelihood is always increasing as β increases, approaching zero as an asymptote. Thus, the likelihood function has no maximum, which implies that the maximum likelihood estimate does not exist.

Figure 3.1 Log-Likelihood as a Function of the Slope under Complete Separation

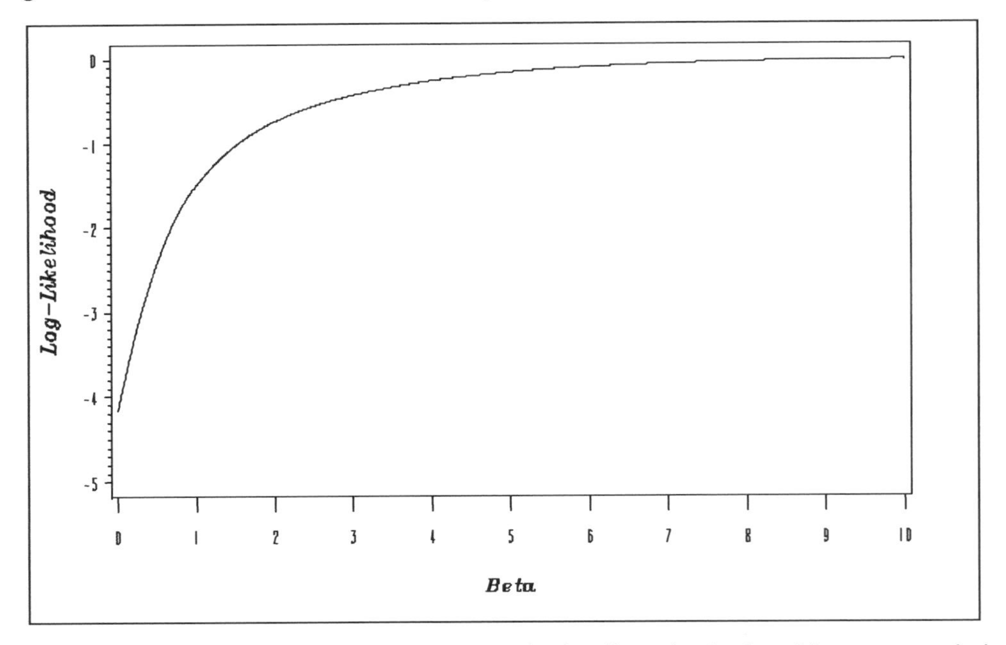

Another peculiarity in Output 3.4 is the disparity in the chi-square statistics for "Testing Global Null Hypothesis: BETA=0." The likelihood ratio chi-square is large (8.3) with a p-value less than .01. The Wald chi-square, on the other hand, is quite small with a p-value of .70. The score statistic is in between. In general, Wald chi-squares tend to be very inaccurate when there is separation in the data, or something approaching separation.

Complete separation is relatively uncommon. Much more common is something called *quasi-complete separation*. We can illustrate this by adding a single observation to the previous data set.

```
DATA quasisep;
  INPUT x y;
  DATALINES;
1 0
2 0
3 0
4 0
4 1
5 1
6 1
;
PROC LOGISTIC;
  MODEL y(EVENT='1') = x;
RUN;
```

The only difference between this and the earlier example is that when x=4, there is one observation that has y=0 and another observation that has y=1. In general, quasi-complete separation occurs whenever there is complete separation except for a single value of the predictor for which both values of the dependent variable occur.

Results in Output 3.4 are quite similar to those in Output 3.3 for complete separation. We again get warning messages, although LOGISTIC tells us that the maximum likelihood *may* not exist rather than *does* not exist. Again, we see a big disparity in the three chi-square statistics for testing the "global null hypothesis." And we get a very large coefficient with a much larger standard error. One difference with complete separation is that the log-likelihood does not go all the way to zero. A graph of the log-likelihood function would look very much like the one in Figure 3.1, but the asymptote for the curve would be something lower than zero.

Output 3.4 PROC LOGISTIC Results under Quasi-Complete Separation

Model Convergence Status
Quasi-complete separation of data points detected.

Warning: The maximum likelihood estimate may not exist.
Warning: The LOGISTIC procedure continues in spite of the above warning. Results shown are based on the last maximum likelihood iteration. Validity of the model fit is questionable.

Model Fit Statistics		
Criterion	Intercept Only	Intercept and Covariates
AIC	11.561	6.774
SC	11.507	6.665
-2 Log L	9.561	2.774

Testing Global Null Hypothesis: BETA=0			
Test	Chi-Square	DF	Pr > ChiSq
Likelihood Ratio	6.7871	1	0.0092
Score	4.2339	1	0.0396
Wald	0.0338	1	0.8542

Analysis of Maximum Likelihood Estimates					
Parameter	DF	Estimate	Standard Error	Wald Chi-Square	Pr > ChiSq
Intercept	1	-33.2785	181.1	0.0337	0.8542
x	1	8.3196	45.2858	0.0338	0.8542

In my experience, the most common cause of quasi-complete separation is a dummy predictor variable that has the following property: at one level of the dummy variable either every case has a 1 on the dependent variable or every case has a 0. To illustrate this, let's restrict the sample of death penalty cases to those with nonblack defendants. Then we'll fit a model with two independent variables, CULP and SERIOUS. We'll treat CULP as a categorical independent variable (with 5 categories) by putting it in a CLASS statement.

```
PROC LOGISTIC DATA=penalty;
  WHERE blackd=0;
  CLASS culp /PARAM=REF;
  MODEL death(EVENT='1') = culp serious;
RUN;
```

LOGISTIC sets up four dummy variables corresponding to the first four categories of CULP. Parameter estimates and associated statistics are shown in Output 3.5. Although not shown in the output, LOGISTIC issued the same warnings for quasi-complete separation that we saw in Output 3.4, and there was again a big disparity between the global likelihood ratio chi-square and the Wald chi-square. What interests us here, however, is the source of the quasi-complete separation.

Output 3.5 LOGISTIC Estimates for Nonblack Defendants Only

Analysis of Maximum Likelihood Estimates						
Parameter		DF	Estimate	Standard Error	Wald Chi-Square	Pr > ChiSq
Intercept		1	3.1273	1.7545	3.1772	0.0747
culp	1	1	-15.5467	148.6	0.0109	0.9167
culp	2	1	-3.8366	1.2267	9.7819	0.0018
culp	3	1	-0.7645	1.1312	0.4568	0.4991
culp	4	1	-0.9595	1.0587	0.8213	0.3648
serious		1	-0.1693	0.1385	1.4950	0.2214

For CULP 1 we see the telltale signs of quasi-complete separation: a coefficient that is very large in magnitude and a standard error that is extremely large, leading to a high *p*-value. To see the cause, take a look at Table 3.1, which shows the cross-classification of DEATH by CULP for the 74 nonblack defendants. Quasi-complete separation occurs because all 30 defendants at level 1 of CULP got life sentences.

Table 3.1 Cross-Classification of Death Penalty by Culpability, Nonblack Defendants

	DEATH		
CULP	0	1	Total
1	30	0	30
2	14	2	16
3	2	4	6
4	3	6	9
5	3	10	13
Total	52	22	74

What can you do about complete or quasi-complete separation? The first task is to figure out which variable (or variables) is causing the problem. As we've seen, they're likely to be the variables with large coefficients and even larger standard errors. To be sure of which variables are causing the problem, you should delete individual variables or sets of variables from the model until you make the problem go away. It's also very helpful to look at cross-classifications of any categorical independent variables with the dependent variable, as we did in Table 3.1. If you find a cell frequency of 0 in any of these tables, you've pinpointed a cause of quasi-complete separation.

After you've found the problem variables, what do you do about them? The answer depends on whether you have complete or quasi-complete separation. If separation is complete, there's no way to get estimates for the coefficients of the other variables as long as the problem variables are in the model. Because the problem variables completely account for the variation in the dependent variable, nothing is left for additional variables to explain. On the other hand, excluding the problem variables from the model isn't very appealing either. The reason they're problematic is that they're *too good* at predicting the dependent variable. Leaving out the strongest predictors can misleadingly suggest that they are unimportant and can also bias the coefficients for remaining variables. If you decide to present results for a model without the problem variables, you should also report that the problem variables perfectly predicted the dependent variable.

If the problem is *quasi*-complete separation, there are additional options to consider.

Recode the problem variables. Consider the model that has quasi-complete separation as shown in Output 3.5. The problem occurred because all persons with CULP=1 got life sentences. One way to eliminate this problem is to take CULP out of the CLASS statement and treat it as a quantitative variable. Of course, this assumes that the effect of CULP is linear, an assumption that may be too restrictive for the data. Another method is to collapse categories 1 and 2 of CULP so that the combined category has 2 death sentences and 44 life sentences:

```
DATA;
  SET penalty;
  IF culp=1 THEN culp=2;
PROC LOGISTIC;
  WHERE blackd=0;
  CLASS culp / PARAM=REF;
  MODEL death(EVENT='1')=culp serious;
RUN;
```

This produced the estimates shown in Output 3.6. Now we get a reasonable estimate for the combined category (CULP 2), one that is highly significant. All warning messages have disappeared.

Output 3.6 Collapsing Categories to Eliminate Quasi-Complete Separation

Analysis of Maximum Likelihood Estimates						
Parameter		DF	Estimate	Standard Error	Wald Chi-Square	Pr > ChiSq
Intercept		1	2.7033	1.6868	2.5682	0.1090
culp	2	1	-4.9386	1.2423	15.8037	<.0001
culp	3	1	-0.7032	1.1205	0.3939	0.5303
culp	4	1	-0.8606	1.0452	0.6781	0.4103
serious		1	-0.1329	0.1340	0.9831	0.3214

Exclude cases from the model. When there is quasi-complete separation, we know that the model perfectly predicts the outcome for some subset of the cases. For example, we just saw that when CULP=1, everyone gets a life sentence. For those individuals, there's nothing left for the other variables to explain, so we might as well restrict the analysis to those cases where CULP > 1. For example,

```
PROC LOGISTIC DATA=penalty;
  WHERE blackd=0 AND culp > 1;
  CLASS culp / PARAM=REF;
  MODEL death(EVENT='1')=culp serious;
RUN;
```

Results are shown in Output 3.7. If you take the approach of excluding cases, you should also report the cross-tabulation of the problem variable with the dependent variable, along with an appropriate test of significance (for example, the Pearson chi-square test for independence).

Output 3.7 Excluding Cases to Eliminate Quasi-Complete Separation

Analysis of Maximum Likelihood Estimates						
Parameter		DF	Estimate	Standard Error	Wald Chi-Square	Pr > ChiSq
Intercept		1	3.1273	1.7545	3.1772	0.0747
culp	2	1	-3.8366	1.2267	9.7819	0.0018
culp	3	1	-0.7645	1.1312	0.4568	0.4991
culp	4	1	-0.9595	1.0587	0.8213	0.3648
serious		1	-0.1693	0.1385	1.4950	0.2214

Retain the model with quasi-complete separation but use likelihood-ratio tests. A third option is to stick with the LOGISTIC model that produced quasi-complete separation. This model controls for the variable that produced the problem, and there is no reason to be suspicious of the results for the *other* variables. As a matter of fact, the results for all the coefficients besides CULP 1 are identical in Output 3.7 (which excluded cases) and in Output 3.5 (which yielded quasi-complete separation). Of course, the coefficient for CULP 1 in Output 3.5 is useless; in research reports, I suggest listing it as ∞. The reported standard error and Wald chi-square for this variable are also useless. Nevertheless, it is still possible to get valid likelihood-ratio tests. PROC GENMOD (which we'll discuss in later chapters) has some easy options for getting likelihood-ratio tests for individual coefficients. PROC LOGISTIC doesn't have those options, but it has a close equivalent—the profile likelihood confidence intervals that we discussed in Section 3.1.

Consider the following program which, as we saw earlier, produced the quasi-complete separation for CULP 1 in Output 3.5:

```
PROC LOGISTIC DATA=penalty;
  WHERE blackd=0;
  CLASS culp /PARAM=REF;
  MODEL death(EVENT='1') = culp serious / CLPARM=PL ALPHA=.01;
RUN;
```

The difference is that there are now two options on the MODEL statement, CLPARM=PL and ALPHA=.01. Together, these options produce 99 percent confidence intervals around the coefficients using the profile likelihood method, as shown in Output 3.8.

Output 3.8 Profile Likelihood Confidence Intervals with Quasi-Complete Separation

Parameter Estimates and Profile-Likelihood Confidence Intervals				
Parameter		**Estimate**	**99% Confidence Limits**	
Intercept		3.1273	-0.9780	8.4133
culp	1	-15.5467	.	-3.2344
culp	2	-3.8366	-7.7102	-1.1199
culp	3	-0.7645	-3.8126	2.4305
culp	4	-0.9595	-3.9167	1.8048
serious		-0.1693	-0.5741	0.1709

For CULP 1, the missing lower confidence limit should be read as minus infinity. Since the 99 percent confidence interval for this variable does not include 0, we can easily reject the null hypothesis that the coefficient is 0 at the .01 level. By contrast, a Wald confidence interval for this variable ranges from -398.3 to 367.2.

Use exact methods. All the models we have estimated so far with PROC LOGISTIC have used maximum likelihood. Although maximum likelihood has many attractive properties, it's important to keep in mind that the standard errors and *p*-values produced by maximum likelihood are large-sample approximations. In small samples, or in cases with separation, the accuracy may not be as good as we would like.

An alternative estimation method that has good small-sample properties is "exact logistic regression." This method can be viewed as a generalization of Fisher's exact test for two-way contingency tables. In this approach, *p*-values are calculated by enumerating all the possible sample outcomes under the null hypothesis. The exact method produces valid *p*-values even in cases of complete or quasi-complete separation. You can implement the method in PROC LOGISTIC with the EXACT statement, as in the following program:

```
PROC LOGISTIC DATA=penalty;
  WHERE blackd=0;
  CLASS culp /PARAM=REF;
  MODEL death(EVENT='1') = culp serious;
  EXACT culp serious / ESTIMATE=BOTH;
RUN;
```

Here the EXACT statement requests exact results for both CULP and SERIOUS. The ESTIMATE=BOTH option requests both parameter estimates and odds ratios. Results are shown in Output 3.9 (not shown are the usual ML estimates that precede these results).

Output 3.9 EXACT Estimates with Quasi-Complete Separation

Conditional Exact Tests				
			p-Value	
Effect	Test	Statistic	Exact	Mid
culp	Score	31.8119	<.0001	<.0001
	Probability	4.53E-10	<.0001	<.0001
serious	Score	1.4073	0.2403	0.2397
	Probability	0.00116	0.6723	0.6718

Exact Parameter Estimates							
Parameter		Estimate		Standard Error	95% Confidence Limits		Two-sided p-Value
culp	1	-4.7520	*	.	-Infinity	-2.8351	<.0001
culp	2	-3.4785		1.1257	-6.6614	-1.0862	0.0010
culp	3	-0.8773		1.1026	-3.5857	2.0040	0.7868
culp	4	-1.0713		1.0389	-3.7447	1.4168	0.5787
serious		-0.1526		0.1311	-0.4268	0.0967	0.2407

Note: * indicates a median unbiased estimate.

Exact Odds Ratios						
Parameter		Estimate		95% Confidence Limits		Two-sided p-Value
culp	1	0.009	*	0	0.059	<.0001
culp	2	0.031		0.001	0.337	0.0010
culp	3	0.416		0.028	7.419	0.7868
culp	4	0.343		0.024	4.124	0.5787
serious		0.858		0.653	1.102	0.2407

Note: * indicates a median unbiased estimate.

The first table in Output 3.9 gives *p*-values for "Conditional Exact Tests." These tests are conditional in the sense that they are based on the conditional distribution of the sufficient statistic for each parameter, conditioning on the sufficient statistics for all the other parameters. (The sufficient statistics are the sums of cross products for each x and the binary dependent variable y.) The tests are exact in the same sense that t-statistics are exact in normal-theory linear regression. That is, they are not large sample approximations, and they give the correct probability of getting a result that is at least as extreme as the one observed in the sample, under the null hypothesis that a variable has no effect. However, there are different ways of defining what is extreme or not extreme. Because of this, we see four different *p*-values reported for each variable (or for each set of design variables in the case of CLASS variables like CULP). We get a "score" test and a "probability" test. And for each of these, we get an "exact" value and a "mid" value. The score test is generally preferred over the probability test because it is more robust to skewness in the distribution of the sufficient statistics. "Mid" is short for "mid-p". It's a correction that is designed to make the test less conservative because of the discreteness of the distribution. In this example, that correction makes little difference.

The title of the next table in Output 3.9, "Exact Parameter Estimates," is somewhat misleading because these estimates are no more exact than conventional maximum likelihood estimates. For variables that do not cause separation, the parameter estimates are "conditional" maximum likelihood estimates. That is, they are estimates that maximize a conditional likelihood that eliminates all other parameters by conditioning on their sufficient statistics. For variables causing separation, the estimates are "median unbiased estimates," which have reasonably good properties in such situations. The "Exact Odds Ratios" in the next table are just the exponentiated values of the parameter estimates. Here we see that the adjusted odds of getting the death penalty at culpability level 1 are estimated to be less than 1 percent of the odds at culpability level 5, the omitted category. With conventional maximum likelihood, we were unable to get a point estimate for this odds ratio.

The *p*-values reported in the tables for parameter estimates and odds ratios are based on likelihood ratio tests. However, the calculation uses the exact distributions of these statistics rather than the usual chi-square approximation. The confidence intervals are also based on the exact distributions. Notice that for the CULP level associated with quasi-complete separation, the lower confidence limit for the parameter estimate is negative infinity and the lower limit for the odds ratio is 0.

Although we have focused on the use of the EXACT statement to handle quasi-complete separation, it's generally useful when sample sizes are small enough to raise doubts about the accuracy of the chi-square approximation. Keep in mind also that when considering sample size, what's important is the number of cases in *the less frequent category* of the dependent variable. Of course, the downside of exact methods is that they can be very computationally demanding. For our death penalty example, conventional maximum likelihood took about a tenth of a second on my laptop. By contrast, exact estimation took about 3.5 *minutes*. Although the theory behind exact methods was developed more than 40 years ago (Cox 1970), it's only in the last decade that these methods have become practical due to improved algorithms and faster computers. PROC LOGISTIC actually has two alternative algorithms that can be chosen with an option on the PROC statement. And there are several additional options that control some of the details of these algorithms.

Use penalized likelihood. One of the easiest and most effective ways of dealing with quasi-complete separation is a method known as penalized likelihood estimation, introduced by Firth (1993) and, hence, often known as the Firth method. It's well known that conventional maximum likelihood estimates may suffer from bias in small samples. The penalized likelihood method is designed to reduce this bias for a wide range of applications of maximum likelihood. Heinze and Schemper (2002) showed that this method is particularly effective in dealing with cases of quasi-complete separation.

Here's how it works. In the Newton-Raphson algorithm (equation 3.7), the vector of first derivatives $\mathbf{U}(\boldsymbol{\beta})$ is modified as follows:

$$\mathbf{U}(\boldsymbol{\beta}) = \frac{\partial \log L}{\partial \boldsymbol{\beta}} = \sum_i \mathbf{x}_i y_i - \sum_i \mathbf{x}_i \hat{y}_i - \sum_i h_i \mathbf{x}_i (.5 - \hat{y}_i)$$

where h_i is the i'th diagonal element of the "hat" matrix

$$\mathbf{H} = \mathbf{W}^{1/2}\mathbf{X}(\mathbf{X'WX})^{-1}\mathbf{X'W}^{1/2} \text{ and } \mathbf{W} = diag\{\hat{y}_i(1-\hat{y}_i\}.$$

Once this replacement is made, the Newton-Raphson algorithm can proceed as usual.

In PROC LOGISTIC, the method is implemented with the FIRTH option on the MODEL statement. Here is the code for the death penalty example:

```
PROC LOGISTIC DATA=penalty;
  WHERE blackd=0;
  CLASS culp /PARAM=REF;
  MODEL death(EVENT='1') = culp serious / FIRTH
    CLPARM=PL;
RUN;
```

Notice that I have requested profile likelihood confidence intervals using the CLPARM option. You should *always* do this when using the Firth method to deal with quasi-complete separation. That's because conventional Wald confidence intervals can be seriously inaccurate in such situations. Wald *p*-values can also be inaccurate for the problematic parameters, so inference should be focused instead on the confidence intervals.

Output 3.10 Penalized Likelihood Estimates with Quasi-Complete Separation

Analysis of Maximum Likelihood Estimates						
Parameter		**DF**	**Estimate**	**Standard Error**	**Wald Chi-Square**	**Pr > ChiSq**
Intercept		1	2.6718	1.5980	2.7956	0.0945
culp	**1**	1	-5.8881	1.7393	11.4602	0.0007
culp	**2**	1	-3.3430	1.1156	8.9794	0.0027
culp	**3**	1	-0.6975	1.0963	0.4047	0.5247
culp	**4**	1	-0.8444	1.0209	0.6841	0.4082
serious		1	-0.1416	0.1265	1.2530	0.2630

Parameter Estimates and Profile-Likelihood Confidence Intervals				
Parameter		**Estimate**	**95% Confidence Limits**	
Intercept		2.6718	-0.2209	6.1591
culp	**1**	-5.8881	-11.0071	-3.1480
culp	**2**	-3.3430	-5.8205	-1.4159
culp	**3**	-0.6975	-2.7747	1.3963
culp	**4**	-0.8444	-2.8643	1.0624
serious		-0.1416	-0.4100	0.0964

Results in Output 3.10 are quite similar to those in Output 3.9 produced by the EXACT option. The penalized likelihood estimate estimate for CULP 1 of -5.8881 is not dissimilar from the median unbiased estimate of -4.7520. Remember that we couldn't get an estimate at all using conventional maximum likelihood. The profile likelihood 95 percent confidence interval of -11.0071 to -3.1480 clearly indicates that this parameter is significantly different from 0. Although the *p*-value for the Wald chi-square (.0007) leads to the same conclusion, you should never trust Wald chi-squares for parameters with non-existent maximum likelihood estimates.

The great thing about the FIRTH method is that the computations are only trivially more difficult than those for conventional maximum likelihood. In fact, for this example, the reported computing times were identical for the two methods. As with the exact method, the FIRTH method is attractive whenever sample sizes are small, even when there is no quasi-complete separation.

3.5 Multicollinearity

One of the nice things about logistic regression is that it's so much like ordinary linear regression analysis. Unfortunately, some of the less pleasant features of linear regression analysis also carry over to logistic regression. One of these is multicollinearity, which occurs when there are strong linear dependencies among the explanatory variables.

For the most part, everything you know about multicollinearity for ordinary regression also applies to logistic regression. The basic point is that, if two or more variables are highly correlated with one another, it's hard to get good estimates of their distinct effects on some dependent variable. Although multicollinearity doesn't bias the coefficients, it does make them more unstable. Standard errors may get large, and variables that appear to have weak effects, individually, may actually have quite strong effects as a group. Fortunately, the consequences of multicollinearity only apply to those variables that are collinear.

How do you diagnose multicollinearity? Examining the correlation matrix produced by PROC CORR may be helpful but is not sufficient. It's quite possible to have data in which no pair of variables has a high correlation, but several variables together may be highly interdependent. Much better diagnostics are produced by PROC REG with the options TOL, VIF, and COLLINOINT. But PROC LOGISTIC doesn't have these options, so what can you do? The thing to remember is that multicollinearity is a property of the explanatory variables, not the dependent variable. So whenever you suspect multicollinearity in a logit model, just estimate the equivalent model in PROC REG and request the collinearity options.

Here's an example using the death penalty data. There are actually two versions of the SERIOUS variable used for the logit model shown in Output 2.2. SERIOUS (the one used before) is based on *rankings* of death penalty cases and SERIOUS2 is based on a 5-point *rating* scale. They have a correlation of .92. Obviously, no sensible person would put both these variables in the same model, but let's see what happens if we do. Output 3.11 shows the results from fitting the model with LOGISTIC. None of the variables is statistically significant even though SERIOUS was highly significant in Output 2.3 On the other hand, the global chi-square of 12.76 is significant at nearly the .01 level. When none of the individual variables is significant but the entire set is significant, multicollinearity is a likely culprit.

Output 3.11 Model with Highly Correlated Explanatory Variables

Testing Global Null Hypothesis: BETA=0			
Test	Chi-Square	DF	Pr > ChiSq
Likelihood Ratio	12.7590	4	0.0125
Score	12.0633	4	0.0169
Wald	11.1765	4	0.0247

Analysis of Maximum Likelihood Estimates					
Parameter	DF	Estimate	Standard Error	Wald Chi-Square	Pr > ChiSq
Intercept	1	-3.1684	0.9867	10.3115	0.0013
blackd	1	0.5854	0.3942	2.2054	0.1375
whitvic	1	0.2842	0.4029	0.4978	0.4805
serious	1	0.0818	0.1532	0.2852	0.5933
serious2	1	0.3967	0.5347	0.5503	0.4582

Let's check it out with PROC REG.

```
PROC REG DATA=penalty;
  MODEL death = blackd whitvic serious serious2 / TOL VIF;
RUN;
```

Output 3.12 gives the collinearity diagnostics. The tolerance is computed by regressing each variable on all the other explanatory variables, calculating the R^2, then subtracting that from 1. Low tolerances correspond to high multicollinearity. While there's no strict cutpoint, I begin to get concerned when I see tolerances below .40. Here we find high tolerances for BLACKD and WHITVIC, but very low tolerances for the two versions of the SERIOUS variable. The variance inflation factor is simply the reciprocal of the tolerance. It tells you how "inflated" the variance of the coefficient is, compared to what it would be if the variable were uncorrelated with any other variable in the model. We see, for example, that the variance of the SERIOUS1 coefficient is 7 times what it would be in the absence of collinearity, implying that its standard error is $\sqrt{7} = 2.6$ times as large.

Output 3.12 Multicollinearity Diagnostics Obtained from PROC REG

Parameter Estimates							
Variable	DF	Parameter Estimate	Standard Error	t Value	Pr > |t|	Tolerance	Variance Inflation
Intercept	1	-0.14164	0.18229	-0.78	0.4384	.	0
blackd	1	0.12093	0.08242	1.47	0.1445	0.85428	1.17058
whitvic	1	0.05739	0.08451	0.68	0.4982	0.84548	1.18276
serious	1	0.01924	0.03165	0.61	0.5442	0.14290	6.99788
serious2	1	0.07044	0.10759	0.65	0.5137	0.14387	6.95081

This approach to diagnosis should be entirely satisfactory in the vast majority of cases, but occasionally it can miss serious multicollinearity (Davis et al. 1986). That's because the linear combinations should ideally be adjusted by the weight matrix used in the maximum likelihood algorithm. If you really want to do it right, here's how.

```
PROC LOGISTIC DATA=penalty;
  MODEL death(EVENT='1') = blackd whitvic serious serious2;
  OUTPUT OUT=a PRED=phat;
DATA b;
  SET a;
  w = phat*(1-phat);
PROC REG DATA=b;
  WEIGHT w;
  MODEL death = blackd whitvic serious1 serious2 / TOL VIF;
RUN;
```

The OUTPUT statement in LOGISTIC creates a new data set that contains all the variables in the MODEL statement plus the variable PHAT that contains the predicted probabilities of the dependent variable. These predicted values are then used in a DATA step to construct the weight variable W. Finally, a weighted least squares regression is performed, using W as the weight variable. For these data, the collinearity diagnostics were only trivially different from those shown in Output 3.12, so I won't bother displaying them.

Diagnosis of multicollinearity is easy. The big problem is what to do about it. The range of solutions available for logistic regression is pretty much the same as for linear regression, such as dropping variables, combining variables into an index, and testing hypotheses about sets of variables. Usually, none of the potential fix-ups is very satisfying.

For a detailed discussion of how to deal with multicollinearity in linear models see Fox (1991).

3.6 Goodness-of-Fit Statistics

The LOGISTIC procedure can produce many different statistics to help you evaluate your models. We've already examined some of the default statistics, notably the global chi-square tests produced by LOGISTIC and the AIC and SC measures of relative fit (Section 2.7). In this section, we'll look at some of the optional statistics.

It's essential to keep in mind that there are many ways of approaching model evaluation, and statistics that appear to be similar may actually answer quite different questions. For example, LOGISTIC's global chi-square addresses the question, "Is this model better than nothing?" A significant chi-square signals a "yes" answer, suggesting that the model is acceptable. By contrast, in this section we will examine the deviance chi-square, which answers the question, "Is there a better model than this one?" Again, a significant chi-square corresponds to a "yes" answer, but that leads to *rejection* of the model.

The deviance is often described as a goodness-of-fit statistic. Such statistics implicitly involve a comparison between the model of interest and a "maximal" model that is more complex. The maximal model always fits better than the model of interest—the question is whether the difference in fit could be explained by chance.

In calculating the deviance for a logistic model, the maximal model is often referred to as the *saturated* model. A saturated model has one parameter for every predicted probability and therefore produces a perfect fit to the data. As a likelihood ratio statistic, the deviance is equal to twice the positive difference between the log-likelihood for the fitted model and the log-likelihood for the saturated model. With individual-level data, the log-likelihood for the saturated model is necessarily 0, so the deviance is just –2 times the log-likelihood for the fitted model.

Unfortunately, with individual-level data the deviance doesn't have a chi-square distribution, for two reasons:

- The number of parameters in the saturated models increase with sample size, thereby violating a condition of asymptotic theory.
- The predicted frequencies for each observation are small (between 0 and 1).

As a result, it's incorrect to compute or interpret a *p*-value for the deviance when working with individual-level data.

If the number of explanatory variables is small and each variable has a small number of values, you can use the AGGREGATE and SCALE options in LOGISTIC to get a deviance that *does* have a chi-square distribution. Consider the model we used in Section 3.2 as an example of computing confidence intervals. Two of the explanatory variables WHITVIC and BLACKD are dichotomous. The other variable CULP has five values. Here's how to get the deviance for the model using PROC LOGISTIC:

```
PROC LOGISTIC DATA=penalty;
  MODEL death(EVENT='1') = blackd whitvic culp / AGGREGATE
     SCALE=NONE;
RUN;
```

The AGGREGATE option tells LOGISTIC to aggregate the data over the various levels of the predictor variables. The SCALE option requests goodness-of-fit statistics, but specifying SCALE=NONE tells LOGISTIC *not* to adjust the goodness-of-fit statistics for overdispersion, a topic we'll examine in Chapter 4.

These options don't change any of the standard output—all they do is produce the new statistics shown in Output 3.13. The *p*-value for the new deviance is .81 indicating a very good fit.

Output 3.13 Statistics from the AGGREGATE and SCALE Options

Deviance and Pearson Goodness-of-Fit Statistics				
Criterion	**Value**	**DF**	**Value/DF**	**Pr > ChiSq**
Deviance	10.0834	15	0.6722	0.8145
Pearson	8.4674	15	0.5645	0.9037

How is this deviance calculated? Under the AGGREGATE option, LOGISTIC forms the cross-classification of all levels of all the explanatory variables. In this case, we get a $2 \times 2 \times 5$ table, but one of those 20 combinations has no cases, leaving us with 19 "unique profiles." For each of those 19 profiles, the observed and expected frequencies for each of the two outcomes of the dependent variable are calculated based on the fitted model. The deviance is then computed as

$$Deviance = 2\sum_j O_j \log\left(\frac{O_j}{E_j}\right) \tag{3.8}$$

where O_j is the observed frequency and E_j is the expected frequency in cell j. In this example, the sum is taken over all 38 cells (19 profiles by 2 categories of the dependent variable). The 15 degrees of freedom come from the 19 profiles minus the 4 estimated parameters in the model.

What is this statistic testing? As noted earlier, the deviance is implicitly contrasting the fitted model with a saturated model. Here the saturated model has one parameter for each of the 19 profiles, not one parameter for each of the 147 cases. We could fit the saturated model by using the following program:

```
PROC LOGISTIC DATA=penalty;
CLASS culp;
MODEL death(EVENT='1') = blackd whitvic culp blackd*whitvic
blackd*culp whitvic*culp blackd*whitvic*culp ;
RUN;
```

The MODEL statement could also be abbreviated:

```
MODEL death(EVENT='1') = blackd|whitvic|culp;
```

The saturated model treats CULP as a categorical variable and includes all possible interactions among the explanatory variables. If we take the positive difference between the -2 log L for this model and the -2 log L for the model with only main effects and with CULP treated as a quantitative variable, we would get the deviance shown in Output 3.13. Hence, that deviance can be interpreted as a test of the null hypothesis that (a) the effect of CULP is linear *and* (b) all the interactions among the three explanatory variables are 0. It is *not* testing whether the inclusion of other variables could improve the fit.

What about the Pearson chi-square? It's an alternative test of the same null hypothesis, calculated by the following well-known formula:

$$X^2 = \sum_j \frac{(O_j - E_j)^2}{E_j} \tag{3.9}$$

As with the deviance, Pearson's chi-square does not have a chi-square distribution when applied to individual-level data.

While the SCALE and AGGREGATE options are often useful, they don't help if there are many explanatory variables or if some of them are measured on a continuum—

situations that are typical in social science research. In those cases, there will be nearly as many profiles as original observations and nothing is accomplished by aggregating. Neither the deviance nor the Pearson chi-square will have true chi-square distributions. To remedy this deficiency, Hosmer and Lemeshow (2000) proposed a test that has rapidly gained widespread use. It may be implemented in LOGISTIC with the LACKFIT option in the MODEL statement. Let's apply it to the model we've just been evaluating.

```
PROC LOGISTIC DATA=penalty;
  MODEL death(EVENT='1') = blackd whitvic culp / LACKFIT;
RUN;
```

The resulting *p*-value of .82, shown in Output 3.14, suggests that the model fits very well.

Output 3.14 Results from LACKFIT Option in PROC LOGISTIC

Partition for the Hosmer and Lemeshow Test					
		death = 1		death = 0	
Group	Total	Observed	Expected	Observed	Expected
1	6	0	0.11	6	5.89
2	24	0	1.04	24	22.96
3	6	1	0.39	5	5.61
4	22	2	1.98	20	20.02
5	10	1	1.38	9	8.62
6	18	3	3.38	15	14.62
7	16	8	5.56	8	10.44
8	16	10	10.30	6	5.70
9	16	13	13.52	3	2.48
10	13	12	12.34	1	0.66

Hosmer and Lemeshow Goodness-of-Fit Test		
Chi-Square	DF	Pr > ChiSq
4.3925	8	0.8201

The Hosmer-Lemeshow (HL) statistic is calculated in the following way. Based on the estimated model, predicted probabilities are generated for all observations. These are sorted by size, and then grouped into approximately 10 intervals. Within each interval, the expected number of events is obtained by adding up the predicted probabilities. The expected number of non-events is obtained by subtracting the expected number of events from the number of cases in the interval. These expected frequencies are compared with observed frequencies by the conventional Pearson chi-square statistic. The degrees of freedom is the number of intervals minus 2. A high *p*-value, as in this example, indicates that the fitted model cannot be rejected and leads to the conclusion that the model fits well. That is, it can't be significantly improved by adding non-linearities and/or interactions.

The HL statistic has become popular because it fills a major need and has no serious competition. But it's a rather *ad hoc* statistic, and its behavior has not been extensively investigated. Hosmer and Lemeshow (2000) reported simulations showing that the statistic has approximately a chi-square distribution under the null hypothesis that the fitted model is correct. That may be true, but my own simulations suggest that it is not a very powerful test. I generated 100 samples of 500 cases each. The model used to produce the data had two explanatory variables and included main effects and an interaction term,

$$\log\left(\frac{p}{1-p}\right) = \beta_0 + \beta_1 x + \beta_2 z + \beta_3 xz ,$$

with $\beta_0 = 0$, $\beta_1 = 1$, $\beta_2 = 1$, and $\beta_3 = .5$. The variables x and z were drawn from standard normal distributions with a correlation of .50. In each sample, I estimated a logistic model *without* the interaction term and applied the HL test. The model was rejected at the .05 level in only 25% of the samples. In other words, in 75% of the samples the HL test failed to reject a model known to be incorrect. I then fitted a logistic model *with* an interaction term in each of the 100 samples. The Wald test for the interaction was significant at the .05 level in 75% of the samples. While this hardly constitutes a definitive study, it suggests caution in concluding that a model is okay just because the HL test is not significant.

There are other potential problems with the HL test. The choice of 10 groups is completely arbitrary, and some statistical packages (e.g., Stata) allow specification of different numbers of groups. I have seen many examples where slight changes in the number of groups result in radically different *p*-values. I've also seen many cases where the inclusion of an interaction that is highly significant by a likelihood ratio test results in a HL statistic that is substantially worse than one for the model without the interaction. That is *not* appropriate behavior. Similarly, I've seen many examples where the inclusion of an interaction that is clearly not significant (again, by a likelihood ratio test) results in a substantial improvement in fit as evaluated by the HL statistic.

All in all, I can't say that I have a lot of confidence in the HL test. So what can you do to evaluate goodness of fit if you're working with individual-level data that cannot be aggregated? Remember that what the deviance, Pearson chi-square, and HL tests are evaluating is whether you can improve the model by including interactions and non-linearities. In my judgment, there's no good substitute for directly testing those possible interactions and non-linearities. Of course, for many models, there will simply be too many potential interactions and non-linearities to test all the possibilities. So you'll have to be selective, focusing on those variables in which you have the greatest interest, or the greatest suspicion that something more might be needed.

3.7 Statistics Measuring Predictive Power

Another class of statistics describes how well you can predict the dependent variable based on the values of the independent variables. This is a very different criterion from the goodness-of-fit measures that we've just been considering. It's entirely possible to have a model that predicts the dependent variable very well, yet has a terrible fit as evaluated by the deviance or the HL statistic. Nor is it uncommon to have a model that fits well, as judged by either of those goodness-of-fit statistics, yet has very low predictive power.

For least squares linear regression, predictive power is usually measured by the coefficient of determination, commonly known as R^2. Many different R^2 measures have been proposed for logistic regression (for a review, see Veall and Zimmerman (1996) or Tjur (2009)), but LOGISTIC only calculates one of them (as well as an adjusted version). Fortunately, it's the one I like best. It's based on the likelihood ratio chi-square for testing the null hypothesis that all the coefficients are 0 (Cox and Snell 1989), which is the statistic reported by LOGISTIC under the heading "Testing Global Null Hypothesis: BETA=0." If we denote that statistic by L^2 and let n be the sample size, the *generalized* R^2 is

$$R^2 = 1 - \exp\left\{-\frac{L^2}{n}\right\} \tag{3.10}$$

Although this is easy to compute with a hand calculator, LOGISTIC will do it for you if you put the RSQ option on the MODEL statement.

Here's the rationale for the generalized R^2. For an ordinary *linear* regression model, it's possible to calculate a likelihood-ratio chi-square for the hypothesis that all coefficients are 0. That statistic is related to the conventional R^2 by the formula (3.10). In other words, the formula used for logistic regression is identical to the formula that applies to linear regression. In fact, this formula can be used for any regression model estimated by maximum likelihood, including the probit model, the Poisson regression model, and the Cox regression model.

In addition to its appeal as a generalization of the conventional R^2, the generalized R^2 has several things going for it:

- It's based on the quantity being maximized, namely the log-likelihood.
- It's invariant to grouping. You get the same R^2 whether you're working with grouped or individual-level data (as long as cases are only grouped together if they're identical on all the independent variables).
- It's readily obtained with virtually all computer programs because the log-likelihood is nearly always reported by default.
- It never diminishes when you add variables to a model.
- The calculated values are usually quite similar to the R^2 obtained from fitting a linear probability model to dichotomous data by ordinary least squares.

As an example of the last point, the generalized R^2 for the model that has CULP, WHITVIC, and BLACKD is .41. When I fit a linear probability model to these data, the R^2 was .47. Keep in mind that while the generalized R^2 may behave quite similarly to the linear model R^2, it *cannot* be interpreted as a proportion of variance "explained" by the independent variables.

A possible drawback of the generalized R^2 is that its upper bound is less than 1 because the dependent variable is discrete. To fix this, LOGISTIC also reports something labeled the "Max-rescaled Rsquare," which divides the original R^2 by its upper bound (Nagelkerke 1991). For the model just considered, the rescaled R^2 is .57.

The generalized R^2 is *not* the same as McFadden's (1974) R^2 which is popular in econometric literature and used in several software packages. McFadden's R^2 is also based on the log-likelihood but with a somewhat different formula. As in formula (3.10), let L^2 be the likelihood ratio chi-square for test the null hypothesis that all the coefficient are 0, and let ℓ_0 be the log-likelihood for a model with no covariates. Then

$$R_{McF}^{\ 2} = \frac{L^2}{-2\log \ell_0}$$

This is easily calculated from the standard output reported by PROC LOGISTIC. For the model we have just been considering, McFadden's R^2 is .41, the same as the generalized R^2. That won't always be the case, however. Unlike the generalized R^2, McFadden's R^2 does not have an upper bound less than 1.

Both the McFadden R^2 and the generalized R^2 are based on the quantity being maximized, namely the log-likelihood. So ML can be thought of as a method that chooses coefficients to maximize the R^2, at least as defined by these measures. That can be seen as a good thing, but it's not such a good thing if you want to compare the predictive power of a logistic regression model and some other method that's not based on the likelihood function, for example, a neural network model. One way to do that is to use the "model-free" R^2 proposed by Tjur (2009). This is easy to compute and has several attractive properties.

Tjur's R^2 is found by calculating the mean of the predicted values for observations with $y = 1$ and the mean of the predicted values for observations with $y = 0$. The R^2 is just the difference between those two means. Here's a SAS program for the death penalty data.

```
PROC LOGISTIC DATA=penalty;
  MODEL death(EVENT='1')=culp whitvic blackd;
  OUTPUT OUT=a PRED=yhat;
PROC MEANS;
 CLASS death;
 VAR yhat;
RUN;
```

Output 3.15 displays the two means. The difference between them is .48, which is a little bit higher than we got with the other two R^2 measures. The upper bound of Tjur's R^2 is 1.

Output 3.15 Means of Predicted Values by Values of the Dependent Variable

Analysis Variable : yhat Estimated Probability						
death	N Obs	N	Mean	Std Dev	Minimum	Maximum
0	97	97	0.1767130	0.2079205	0.0189454	0.9411508
1	50	50	0.6571852	0.2936428	0.0643906	0.9739092

Another approach to getting a model-free measure of predictive power is to use an ordinal measure of association. PROC LOGISTIC reports four such measures by default whenever you run a binary (or ordinal) logistic regression. Output 3.16 shows the results for the model we've been examining. Four measures of association are shown in the right-hand column. The left-hand column gives the intermediate calculations on which those four statistics are based.

Output 3.16 Ordinal Measures of Association from PROC LOGISTIC

Association of Predicted Probabilities and Observed Responses			
Percent Concordant	88.3	Somers' D	0.800
Percent Discordant	8.4	Gamma	0.827
Percent Tied	3.3	Tau-a	0.361
Pairs	4850	c	0.900

Here's how the measures are calculated. For the 147 observations in the sample, there are 147(146)/2 = 10731 different ways to pair them up (without pairing an observation with itself). Of these, 5881 pairs have either both 1's on the dependent variable or both 0's. We ignore these, leaving 4850 pairs in which one case has a 1 and the other case has a 0. For each pair, we ask the question, "Does the case with a 1 have a higher predicted value (based on the model) than the case with a 0?" If the answer is yes, we call that pair *concordant*. If no, the pair is *discordant*. If the two cases have the same predicted value, we call it a tie. Let C be the number of concordant pairs, D the number of discordant pairs, T the number of ties, and N the total number of pairs (before eliminating any). The four measures of association are then defined as follows:

$$Tau-a = \frac{C-D}{N}$$

$$Gamma = \frac{C-D}{C+D}$$

$$Somer's\ D = \frac{C-D}{C+D+T}$$

$$c = .5\ (1 + Somer's\ D).$$

All four measures vary between 0 and 1, with large values corresponding to stronger associations between the predicted and observed values. Of the four statistics, Tau-a tends to be closest to the generalized R^2. On the other hand, the c statistic has become very popular because it corresponds to the area under the ROC curve, which is the topic of the next section.

3.8 ROC Curves

Another approach to evaluating the predictive power of models for binary outcomes is the ROC curve. ROC is short for *Receiver Operating Characteristic*. That name stems from the fact that the methodology was first developed during World War II to evaluate the performance of radar receivers in the detection of enemy aircraft.

To understand the ROC curve, you must first understand classification tables. As usual, our dependent variable y has values of 1 or 0. Let $\hat{p}_i$ be the predicted probability that $y_i = 1$ for individual i, based on some model that we have estimated. If we want to use the predicted probabilities to generate actual predictions of whether or not $y = 1$, we need some cutpoint value. A natural choice would be .5. If $\hat{p}_i \geq .5$, we predict $y_i = 1$. If $\hat{p}_i < .5$, we predict $y_i = 0$.

For the logistic regression model that we estimated in the last section (with DEATH as the dependent variable and CULP, BLACKD, and WHITVIC as predictors), we can then construct the following classification table:

	$\hat{p}_i \geq .5$	$\hat{p}_i < .5$	Total
$y_i = 1$	35	15	50
$y_i = 0$	10	87	97

Thus, 35 persons who got the death penalty are correctly predicted but 15 are not. Of those who did not get the death penalty, 87 are correctly predicted but 10 are not. This seems pretty accurate, but we want a more precise way of quantifying that accuracy. One way is to compute the overall proportion of predictions that are correct: (35+87)/(50+97) = .83. So 83 percent of our predictions are correct which, again, seems quite good. But the overall proportion correct can be quite misleading. Suppose that a data set has 100 events and 900 non-events. A model with *no* predictors will generate predicted values that are all .10, and thus all the cases would be predicted as non-events. This model would be right 90 percent of the time, but clearly that's not a good indication of the model's predictive accuracy.

To do better, we first define *sensitivity* and *specificity*. Sensitivity is the proportion of events that are correctly predicted, in this case 35/50 = .70. Specificity is the proportion of *non*-events that are correctly predicted, i.e., 87/97 = .90. The ideal is for both of these proportions to be high. In our hypothetical example with 100 events, 900 non-events and no predictors, the specificity would be 1.00 but the sensitivity would be 0, which is a clearly unacceptable result. Keep in mind that, in many applications, the choice of which of the two outcomes gets labeled as an event is completely arbitrary. And that implies that which of the two proportions gets labeled as sensitivity and which is labeled as specificity is also arbitrary.

I produced the numbers in the classification table above with PROC LOGISTIC by including the options CTABLE and PPROB=.5 on the MODEL statement. Although .5 might seem like the most natural cutpoint for prediction, we could certainly try others. In fact, if you use the CTABLE option without the PPROB option, you get classification tables for a wide range of possible cutpoint values, as shown in Output 3.17 where the cutpoint probabilities (in the first column) range from 0 to .98, incrementing by .02 for each table. Each row corresponds to a different 2 × 2 table. The two frequencies under the "Correct" heading should be unambiguous. Under the "Incorrect" heading, an incorrect "Event" is an observation that is predicted to be an event but was actually a non-event. And similarly, an incorrect "Non-Event" is an observation that is predicted to be a non-event but actually was an event. The column labeled "Correct" is the overall percentage of predictions that were correct. Interestingly, this reaches its highest value of 84.4 percent for cutpoints that range from .28 to .34.

The next two columns report the sensitivity and specificity for each cutpoint, reported as percentages. The most striking thing about these columns is the inverse relationship between specificity and sensitivity. The closer the cutpoint is to zero, the higher probability that an observation will be predicted to be an event. So most events will be predicted to be events, and sensitivity will be high. But most non-events will also be predicted to be events, and specificity will be low. When the cutpoint is high, you get the opposite pattern. A much larger fraction of the observations are predicted to be non-events, so sensitivity is low but specificity is high.

Output 3.17 Classification Tables for Varying Cutpoints

Classification Table									
	Correct		Incorrect		Percentages				
Prob Level	Event	Non-Event	Event	Non-Event	Correct	Sensi-tivity	Speci-ficity	False POS	False NEG
0.000	50	0	97	0	34.0	100.0	0.0	66.0	.
0.020	50	6	91	0	38.1	100.0	6.2	64.5	0.0
0.040	50	6	91	0	38.1	100.0	6.2	64.5	0.0
0.060	49	30	67	1	53.7	98.0	30.9	57.8	3.2
0.080	47	35	62	3	55.8	94.0	36.1	56.9	7.9
0.100	47	55	42	3	69.4	94.0	56.7	47.2	5.2
0.120	47	55	42	3	69.4	94.0	56.7	47.2	5.2
0.140	46	55	42	4	68.7	92.0	56.7	47.7	6.8
0.160	46	64	33	4	74.8	92.0	66.0	41.8	5.9
0.180	43	64	33	7	72.8	86.0	66.0	43.4	9.9
0.200	43	79	18	7	83.0	86.0	81.4	29.5	8.1
0.220	43	79	18	7	83.0	86.0	81.4	29.5	8.1
0.240	41	79	18	9	81.6	82.0	81.4	30.5	10.2
0.260	41	79	18	9	81.6	82.0	81.4	30.5	10.2
0.280	41	83	14	9	84.4	82.0	85.6	25.5	9.8
0.300	41	83	14	9	84.4	82.0	85.6	25.5	9.8
0.320	41	83	14	9	84.4	82.0	85.6	25.5	9.8
0.340	41	83	14	9	84.4	82.0	85.6	25.5	9.8
0.360	37	83	14	13	81.6	74.0	85.6	27.5	13.5
0.380	37	85	12	13	83.0	74.0	87.6	24.5	13.3
0.400	37	85	12	13	83.0	74.0	87.6	24.5	13.3
0.420	37	85	12	13	83.0	74.0	87.6	24.5	13.3
0.440	35	85	12	15	81.6	70.0	87.6	25.5	15.0
0.460	35	85	12	15	81.6	70.0	87.6	25.5	15.0
0.480	35	87	10	15	83.0	70.0	89.7	22.2	14.7
0.500	35	87	10	15	83.0	70.0	89.7	22.2	14.7
0.520	35	88	9	15	83.7	70.0	90.7	20.5	14.6
0.540	33	88	9	17	82.3	66.0	90.7	21.4	16.2
0.560	33	88	9	17	82.3	66.0	90.7	21.4	16.2
0.580	33	88	9	17	82.3	66.0	90.7	21.4	16.2
0.600	33	90	7	17	83.7	66.0	92.8	17.5	15.9
0.620	33	90	7	17	83.7	66.0	92.8	17.5	15.9
0.640	33	90	7	17	83.7	66.0	92.8	17.5	15.9

Classification Table									
	Correct		Incorrect		Percentages				
Prob Level	Event	Non-Event	Event	Non-Event	Correct	Sensi-tivity	Speci-ficity	False POS	False NEG
0.660	27	90	7	23	79.6	54.0	92.8	20.6	20.4
0.680	27	90	7	23	79.6	54.0	92.8	20.6	20.4
0.700	27	92	5	23	81.0	54.0	94.8	15.6	20.0
0.720	27	92	5	23	81.0	54.0	94.8	15.6	20.0
0.740	23	92	5	27	78.2	46.0	94.8	17.9	22.7
0.760	23	92	5	27	78.2	46.0	94.8	17.9	22.7
0.780	23	92	5	27	78.2	46.0	94.8	17.9	22.7
0.800	23	93	4	27	78.9	46.0	95.9	14.8	22.5
0.820	20	94	3	30	77.6	40.0	96.9	13.0	24.2
0.840	20	94	3	30	77.6	40.0	96.9	13.0	24.2
0.860	20	94	3	30	77.6	40.0	96.9	13.0	24.2
0.880	12	94	3	38	72.1	24.0	96.9	20.0	28.8
0.900	12	94	3	38	72.1	24.0	96.9	20.0	28.8
0.920	11	96	1	39	72.8	22.0	99.0	8.3	28.9
0.940	4	96	1	46	68.0	8.0	99.0	20.0	32.4
0.960	4	96	1	46	68.0	8.0	99.0	20.0	32.4
0.980	0	97	0	50	66.0	0.0	100.0	.	34.0

Before going any further, I need to make a brief technical digression. The predicted probabilities used in the calculations reported in Output 3.17 are not the usual predicted values that you might produce using the OUTPUT statement in PROC LOGISTIC. Instead, they are "bias corrected" predicted values. The rationale is that if you include a particular observation in the sample used to estimate the logistic regression coefficients, you artificially make it more likely that the model will correctly predict that observation. A better method would be to delete the observation from the sample, re-estimate the logistic regression, and then use the results of that regression to generate a predicted probability for deleted observation. PROC LOGISTIC doesn't actually do that because it would have to run a separate logistic regression for every case in the sample. But what it does is a very close approximation.

Returning to Output 3.17, one might ask, "What's the best cutpoint value?" Some authors recommend choosing a cutpoint that produces approximately equal values of sensitivity and specificity. For Output 3.17, that result occurs with a cutpoint around .25. But, ideally, the choice would depend on your assessment of the relative costs of the two kinds of errors. Failing to detect a disease might be regarded as a substantially more costly error than diagnosing someone with the disease who doesn't really have it.

Now we are finally ready to talk about ROC curves. The ROC curve gives us a way of graphically summarizing the information in Output 3.17. It also provides the basis for calculating a single statistic that assesses the predictive power of the model and does not depend on the cutpoint value. The ROC curve is simply a graph with sensitivity on the vertical axis and 1 minus specificity on the horizontal axis, both of which increase as the cutpoint decreases from 1 to 0. PROC LOGISTIC can produce the curve with the following code:

```
PROC LOGISTIC DATA=penalty PLOTS(ONLY)=ROC;
  MODEL death(EVENT='1')=blackd whitvic culp ;
RUN;
```

The option ONLY after the PLOTS option is included to suppress the production of influence plots that would otherwise appear by default. We'll discuss influence plots in the next section.

The resulting ROC curve is shown in Output 3.18. The 45-degree line represents the expected ROC curve for a model with an intercept only, that is, one with no predictive power. The more the curve departs from the 45-degree line, the greater the predictive power. The standard statistic for summarizing that departure is the area under the curve, which here is reported as .8998. As noted earlier, this is the same as the *c* statistic that is reported in the "Association of Predicted Probabilities and Observed Responses" table. (There may be a slight difference because different methods are used to approximate the area.) So if all you want is the area under the curve, there is no need to request the curve itself. One technical note: The predicted probabilities used in the construction of the ROC curve are *not* biased corrected like the ones used to produce Output 3.17.

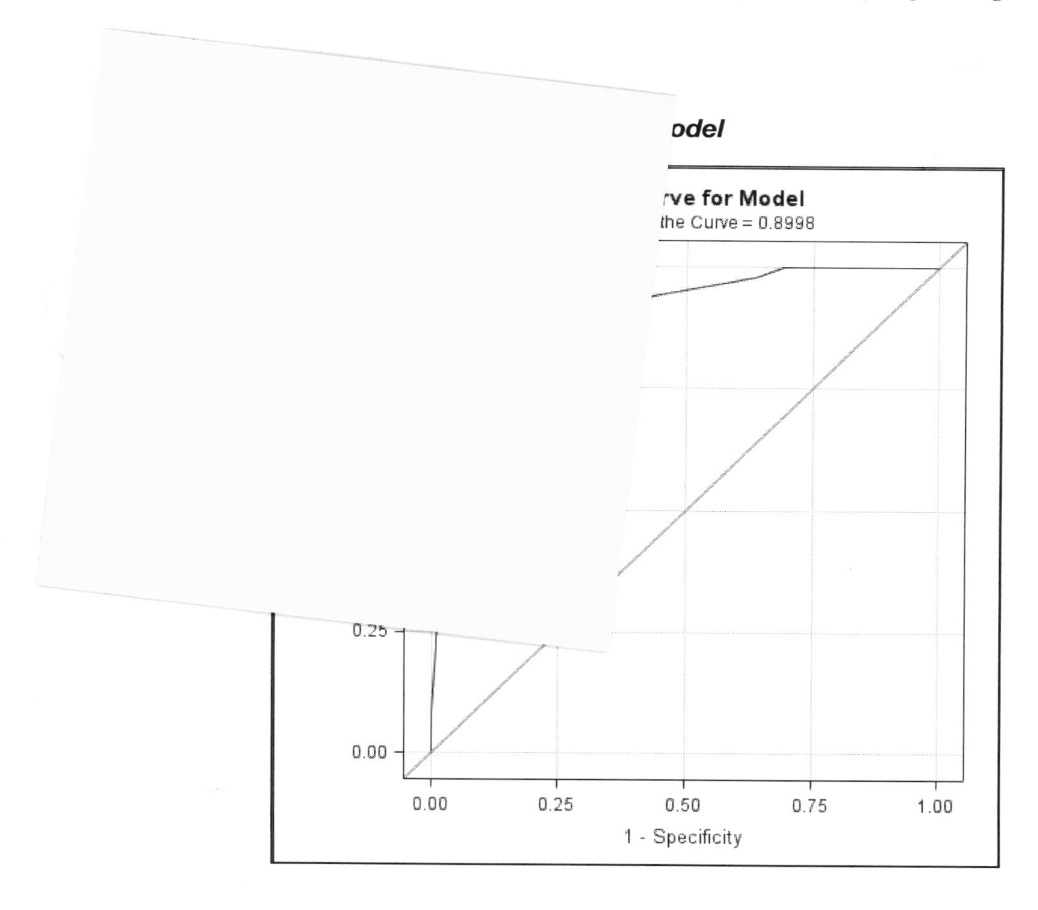

There are lots of options for ROC curves, only a few of which we can examine here. One handy option lets you put the cutpoints on the curve. That's accomplished by changing the PLOTS option to PLOTS(ONLY)=ROC(ID=CUTPOINT), yielding the graph in Output 3.19.

Output 3.19 ROC Curve with Cutpoints Added

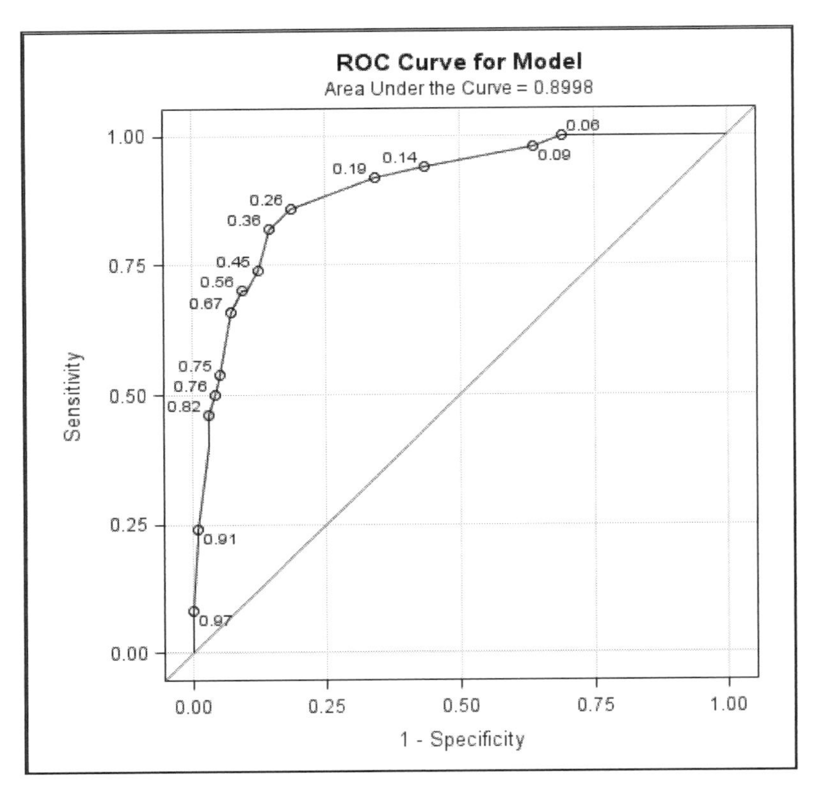

Another attractive feature is the ability to compare the ROC curves and *c*-statistics for different models. In the following code, I request ROC curves and *c*-statistics for the basic model and for three different submodels, each omitting one of the model predictors:

```
PROC LOGISTIC DATA=penalty;
  MODEL death(EVENT='1')=blackd whitvic culp;
  ROC 'omit culp' blackd whitvic;
  ROC 'omit blackd' whitvic culp;
  ROC 'omit whitvic' blackd culp;
  ROCCONTRAST / ESTIMATE=ALLPAIRS;
RUN;
```

The label (enclosed in single quotes) on each ROC statement is optional, but usually helpful in interpreting the output. The three ROC statements produce separate ROC graphs for each model (not shown) and a final graph that displays all the curves in one graph (shown in Output 3.20). Note that when the program contains ROC statements, the PLOTS option is not necessary on the PROC statement. However, you may still want to use it to control some of the plotting details (like the inclusion of cutpoints).

Output 3.20 ROC Curves for Comparing Different Models

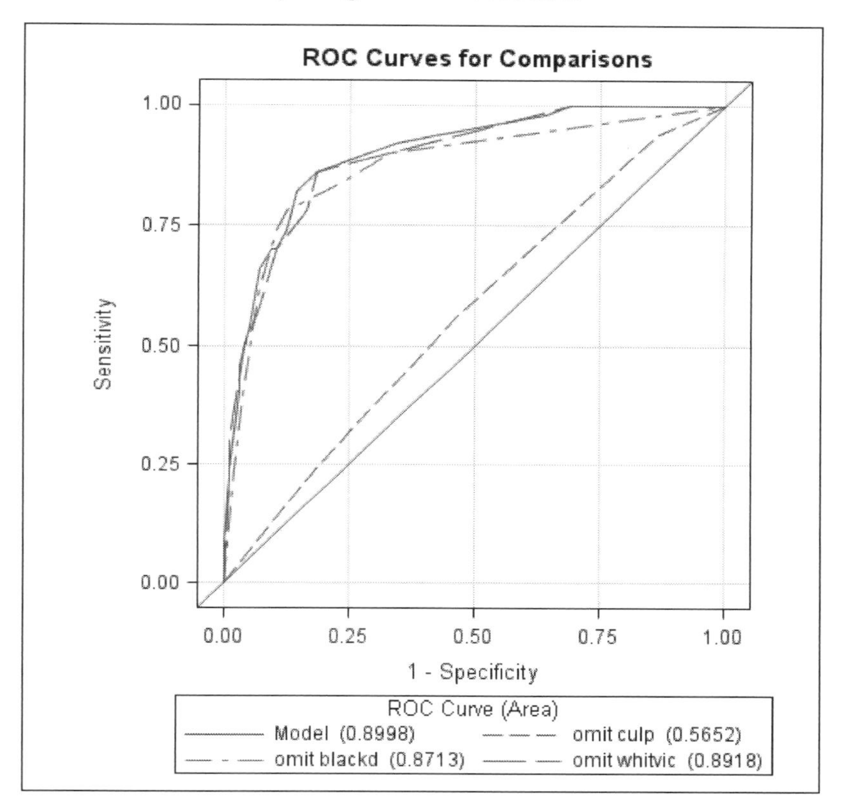

Besides the graphs, we also get the table in Output 3.21, which reports various ordinal measures of association between observed and predicted values for the three models specified in ROC statements, as well as for the original model. "Area" is, of course, the area under the ROC curve. For that statistic we also get standard errors and 95 percent confidence intervals. From both the graph and the table, we see clearly that the variable CULP has a much stronger impact on the association measures than the other two variables.

Output 3.21 Association Statistics for Three Different Models

ROC Association Statistics							
	Mann-Whitney						
ROC Model	Area	Standard Error	95% Wald Confidence Limits		Somers' D (Gini)	Gamma	Tau-a
Model	0.8998	0.0264	0.8481	0.9515	0.7996	0.8272	0.3614
omit culp	0.5652	0.0469	0.4732	0.6571	0.1303	0.1847	0.0589
omit blackd	0.8713	0.0331	0.8065	0.9362	0.7427	0.7934	0.3357
omit whitvic	0.8918	0.0269	0.8390	0.9445	0.7835	0.8363	0.3541

The ROCCONTRAST statement enables us to test for differences in the area under the curve (*c*-statistic) for the different models, as shown in Output 3.22. We first get an overall test of the null hypothesis that the *c*-statistic is the same across all four models (the main model with all three covariates and the three submodels that omit one covariate each). Clearly that hypothesis must be rejected. Then, because I included the ESTIMATE=ALLPAIRS option, we get the difference in the *c*-statistics for each pair of models, along with its standard error, 95 percent confidence interval, and *p*-value for testing the null hypothesis that the true difference is 0. The only significant differences are between models that include CULP and models that exclude it.

Output 3.22 Tables Produced by ROCCONTRAST Statement

ROC Contrast Test Results			
Contrast	DF	Chi-Square	Pr > ChiSq
Reference = Model	3	95.6616	<.0001

ROC Contrast Estimation and Testing Results by Row						
Contrast	Estimate	Standard Error	95% Wald Confidence Limits		Chi-Square	Pr > ChiSq
Model - omit culp	0.3346	0.0513	0.2341	0.4352	42.5659	<.0001
Model - omit blackd	0.0285	0.0179	-0.00656	0.0635	2.5366	0.1112
Model - omit whitvic	0.00804	0.00921	-0.0100	0.0261	0.7625	0.3826
omit culp - omit blackd	-0.3062	0.0637	-0.4310	-0.1813	23.1091	<.0001
omit culp - omit whitvic	-0.3266	0.0510	-0.4265	-0.2267	41.0821	<.0001
omit blackd - omit whitvic	-0.0204	0.0225	-0.0645	0.0237	0.8228	0.3644

3.9 Predicted Values, Residuals, and Influence Statistics

Next, we turn to diagnostic statistics that are computed for each individual observation. PROC LOGISTIC can produce a large number of case-wise statistics with the OUTPUT statement, which writes selected diagnostic statistics to a SAS data set. Here are some of the statistics that can be selected:

- **Linear predictor**—Predicted log-odds for each case. In matrix notation, this is $\mathbf{x}\boldsymbol{\beta}$, so it's commonly referred to as XBETA.
- **Standard error of linear predictor**—Used in generating confidence intervals.
- **Predicted values**—Predicted probability of the event, based on the estimated model and values of the explanatory variables. For grouped data, this is the expected number of events.
- **Confidence intervals for predicted values**—Confidence intervals are first calculated for the linear predictor by adding and subtracting an appropriate multiple of the standard error. Then, to get confidence intervals around the predicted values, the upper and lower bounds on the linear predictor are substituted into $1/(1+e^{-x})$, where x is either an upper or a lower bound.
- **Deviance residuals**—Contribution of each observation to the deviance chi-square.
- **Pearson residuals**—Contribution of each observation to the Pearson chi-square.

For the last two statistics, standardized versions are available in SAS 9.3 and later.

Obviously, the residuals may be useful in determining which observations are poorly fit by the model. But since the dependent variable can only take on values of 0 or 1, the utility of residuals is somewhat limited. A high residual would mean that the individual had the event, even though the predicted probability of the event was low. But we know that low probability events do happen sometimes, so this doesn't necessarily mean a failure of the model, or that something is amiss with that individual. For example, using the model for the death penalty data with CULP, WHITVIC, and BLACKD as predictors, person 10 had both the largest deviance residual (-2.38) and the largest Pearson residual (-4.00). That person did *not* get the death penalty even though his predicted probability was .95.

The OUTPUT statement can also produce several statistics that are designed to measure the *influence* of each observation. Basically, influence statistics tell you how much some feature of the model changes when a particular observation is deleted from the data set. For linear models estimated with PROC REG, influence statistics are exact. Exact computations in logistic models would be very intensive, however, so only approximations are used. The available influence statistics in LOGISTIC are

- **DFBETAS**—These statistics tell you how much each regression coefficient changes when a particular observation is deleted. The actual change is divided by the standard error of the coefficient.
- **DIFDEV**—Change in deviance with deletion of the observation.
- **DIFCHISQ**—Change in Pearson chi-square with deletion of the observation.
- **C and CBAR**—Measures of overall change in regression coefficients, analogous to Cook's distance in linear regression.
- **LEVERAGE**—Measures how extreme the observation is in the space of the explanatory variables. The leverage is the diagonal of the "hat" matrix.

You can get any of these statistics by specifying the appropriate option on the OUTPUT statement. But you may find it more informative to request ODS graphics using the PLOTS option on the PROC statement. The many plots produced by this option come in three varieties that differ according to which variable is plotted on the horizontal axis. For INFLUENCE and DFBETAS plots, the horizontal axis is just the case number. For PHAT and DPC plots, the horizontal axis is the predicted probability. For LEVERAGE plots, the horizontal axis is the leverage. Here's an example:

```
PROC LOGISTIC DATA=penalty PLOTS(UNPACK LABEL)=
    (INFLUENCE DFBETAS PHAT DPC LEVERAGE);
  MODEL death(EVENT='1')=blackd whitvic culp ;
RUN;
```

This program produces a total of 17 plots, only a few of which will be shown here. The UNPACK option produces a separate graph for each plot, rather than "paneling" several of them into a single graphic. The LABEL option requests that the case number be used as the data point marker in each graph, making it easier to identify which cases are problematic.

Output 3.23 displays one of the seven plots produced by the INFLUENCE option, the change in deviance that would result from the deletion of each of the observations. We see that the largest changes come from observations 10 and 122. Person 10 is the same one that we identified above as having one of the largest deviance and Pearson residuals.

Although not visible in this printing, the observations are color coded in the original graphic, with blue indicating a 1 on the response variable and red indicating a 0.

Output 3.23 Graph of Change in Deviance from Deletion of Each Observation

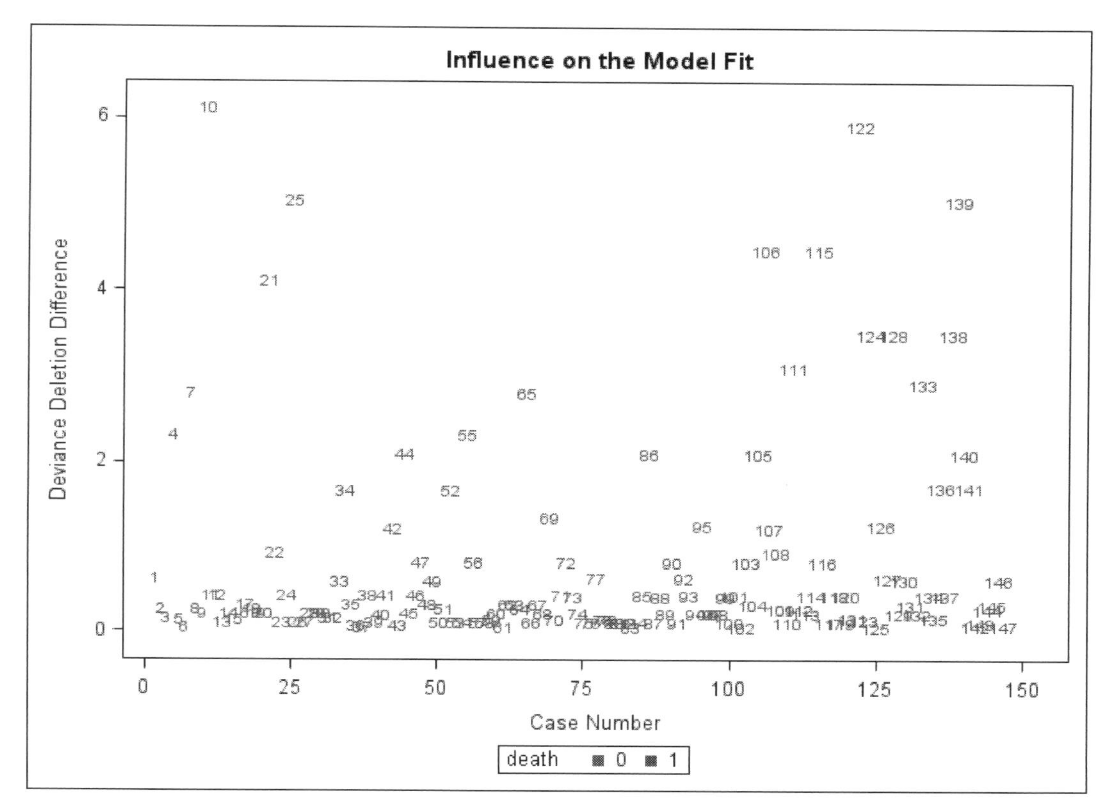

Output 3.24 displays one of the graphs produced by the DFBETAS option. The vertical axis is the change in the coefficient for BLACKD when an observation is deleted (divided by the standard error). The sign of the statistic is the opposite of what you might expect. A positive sign means that the coefficient goes down with deletion of a case, and a negative sign means that the coefficient goes up with case deletion. The most influential observation is clearly person 122 with a DFBETAS value of about -.5. So deleting that case would make the coefficient get larger. A graph like this is produced for each of the predictor variables, as well as for the intercept.

Output 3.24 DFBETAS for the BLACKD Coefficient

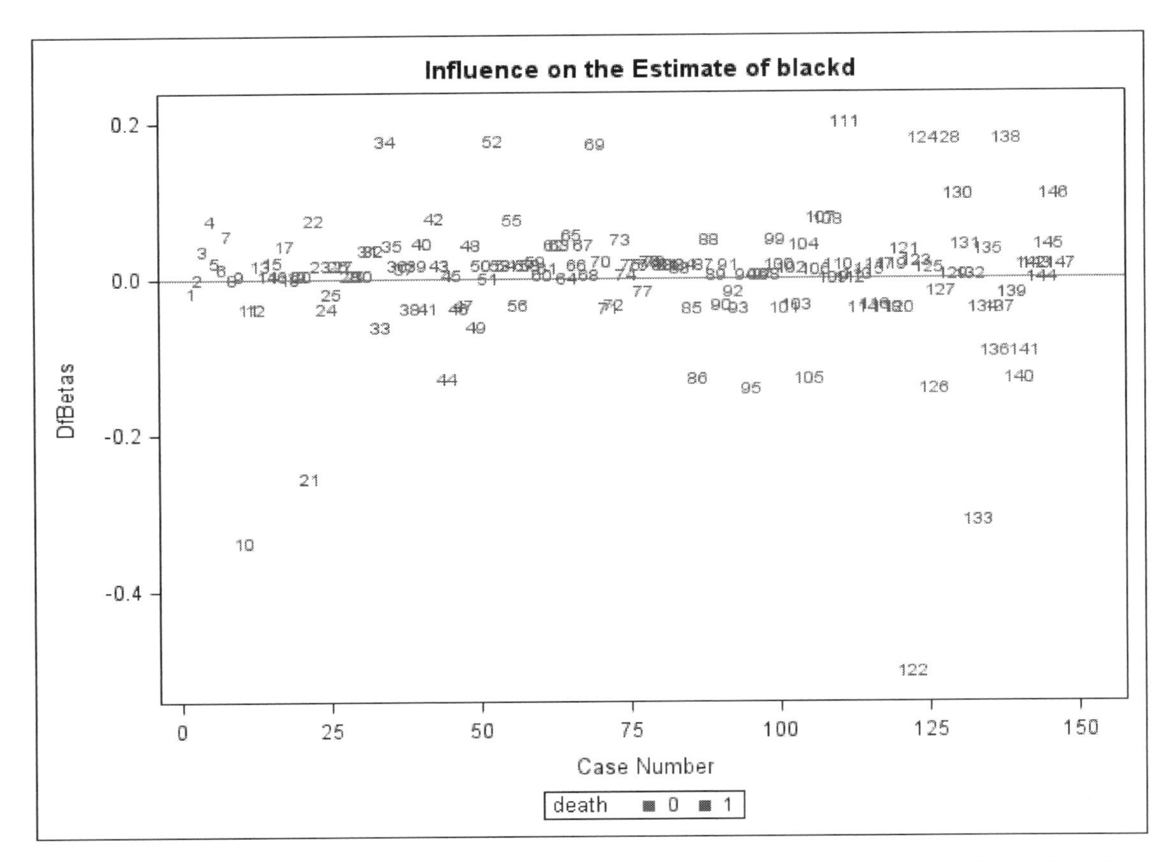

Output 3.25 displays one of the four graphs produced by the PHAT option, which is a plot of the deviance influence statistics (DIFDEV) versus the predicted values. Output 3.25 is fairly typical of this kind of diagnostic plot. The curve sloping upward to the right corresponds to observations that have the dependent variable equal to 0. The curve sloping downward from the left corresponds to observations with the dependent variable equal to 1. According to Hosmer and Lemeshow (2000), who recommend this as one of the more useful diagnostic plots, you should look for points on the graph with high values of the influence statistic, and that are also well separated from other points. The two points in the upper right-hand corner might satisfy that criterion.

Output 3.25 Plot of DIFDEV Statistics versus Predicted Values

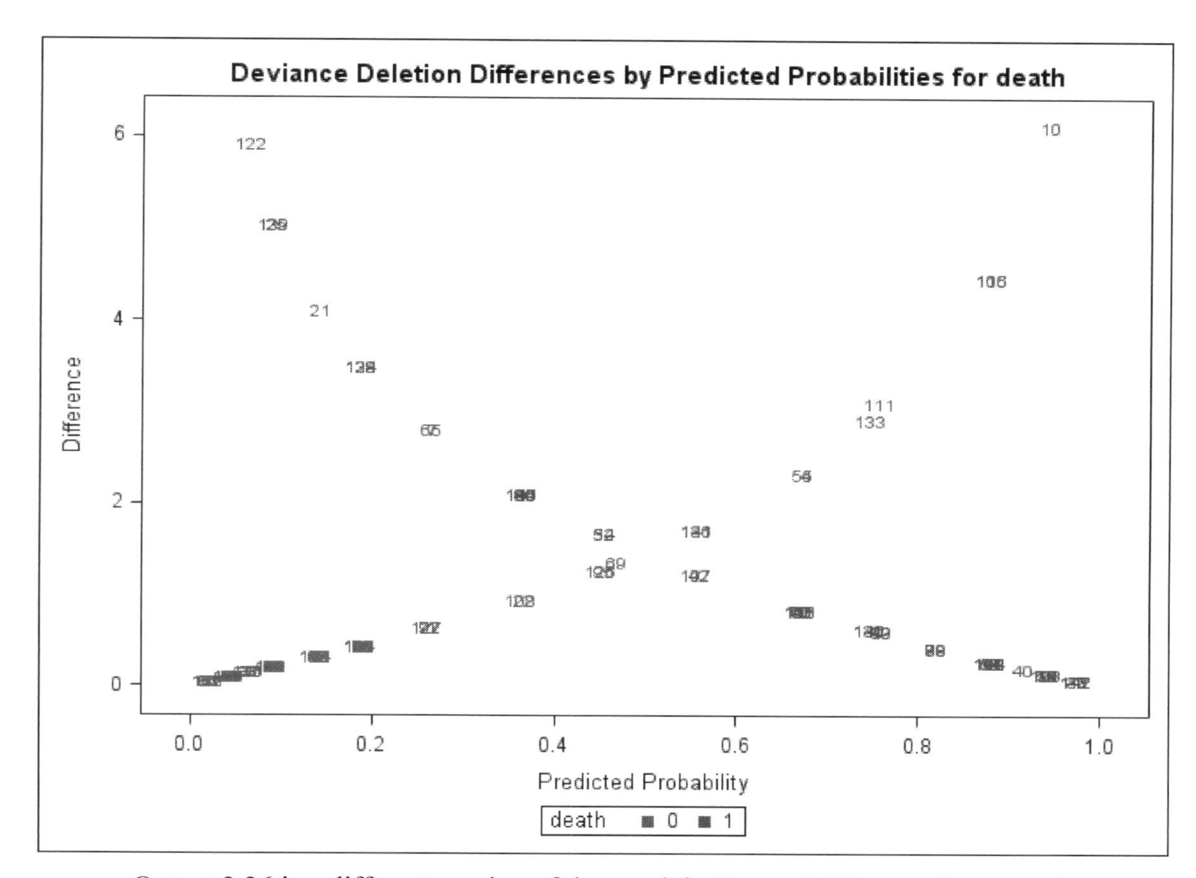

Output 3.26 is a different version of the graph in Output 3.25, one of two graphs produced by the DPC option. This is actually a kind of three-dimensional graph because the observations are color-coded by their values on another influence statistic, the confidence interval displacement statistic C.

Output 3.26 Plot of DIFDEV Statistics versus Predicted Values with Color-Coding by CI Displacement

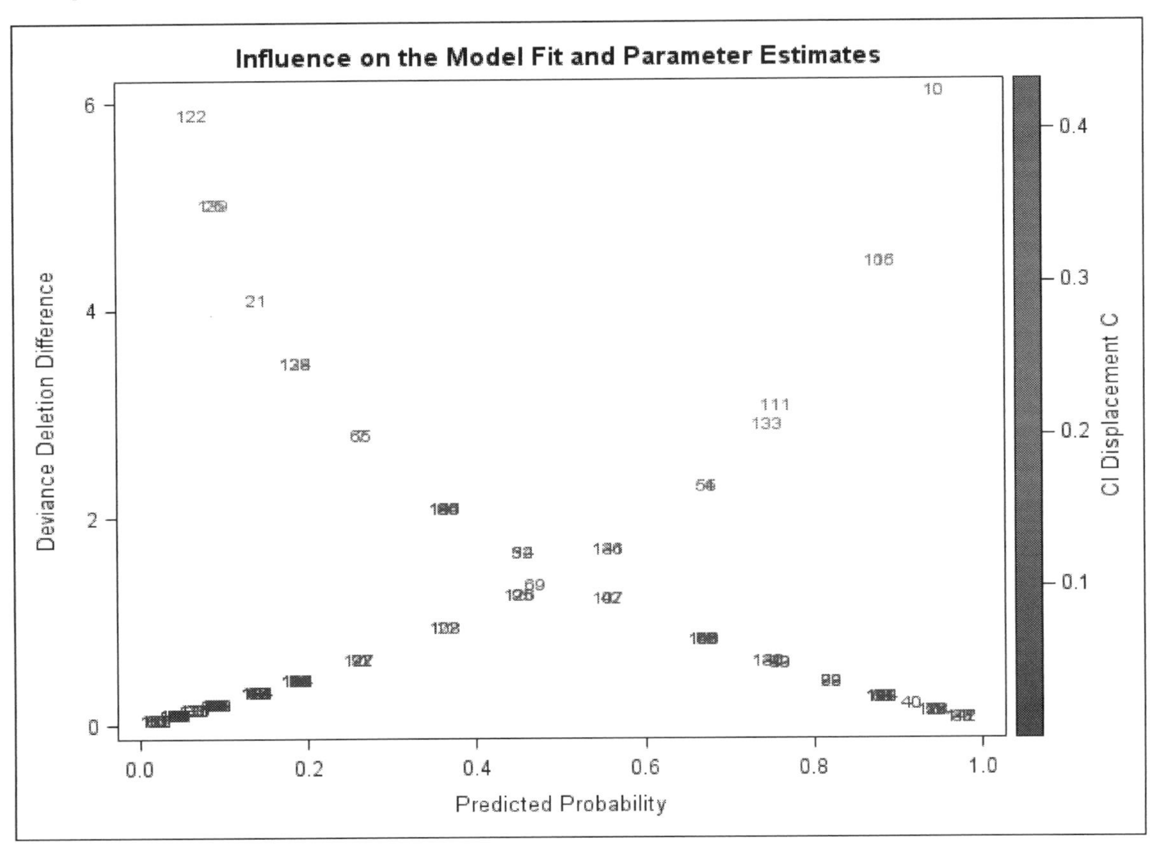

Finally, Output 3.27 displays one of four plots produced by the LEVERAGE option. This plot is the deviance difference statistic with the leverage statistic as the horizontal axis. Concern should be directed toward those observations that are high on both leverage and the deviance statistic, possibly case 111 on this plot.

Output 3.27 Plot of DIFDEV Statistics versus Leverage

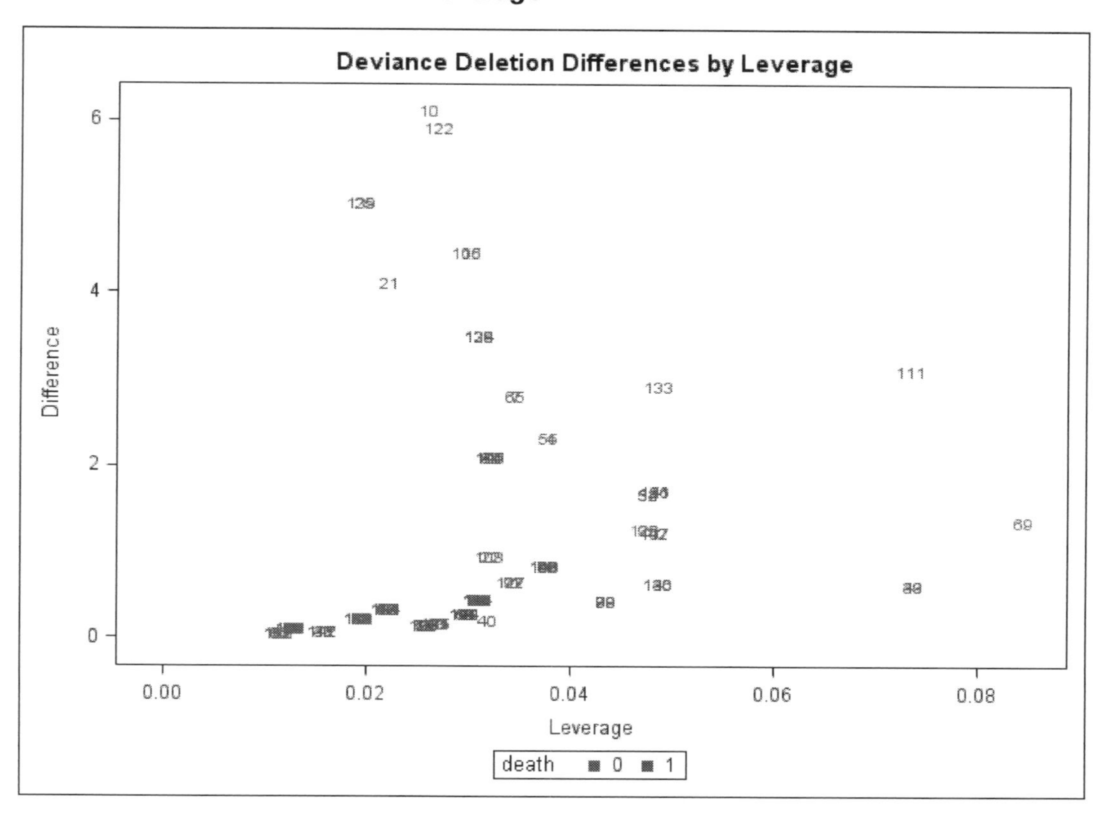

3.10 Latent Variables and Standardized Coefficients

The logistic regression model can be derived by assuming an underlying model for a continuous variable which is then dichotomized. While this rationale is not essential, it can be helpful in understanding several aspects of the model:

- How the model might have arisen in the real world.
- Where standardized coefficients come from.
- How the logit model is related to the probit and complementary log-log models.
- How the model can be generalized to ordinal dependent variables.
- How unobserved heterogeneity can affect logit coefficients.

In this section, we'll examine the first two points, leaving the other three for later.

Here's the idea. Suppose we have a dependent variable z that is measured on a continuous scale. And suppose further that z is related to a set of explanatory variables $x_1, x_2, \ldots, x_k$ by a linear model,

$$z = \alpha_0 + \alpha_1 x_1 + \alpha_2 x_2 + \ldots + \alpha_k x_k + \sigma\varepsilon \qquad (3.11)$$

where ε is a random disturbance term, independent of the x's, and σ is a constant. Now suppose that we can't directly observe z. Instead, we observe y which is equal to 1 if z is greater than some constant μ and is equal to 0 if z is less than or equal to μ. We then ask, how is y related to the x's? The answer depends on the probability distribution of ε.

Suppose ε has a *standard logistic distribution*. That is, ε has a probability density function given by

$$f(\varepsilon) = \frac{e^{\varepsilon}}{\left(1 + e^{\varepsilon}\right)^2} \qquad (3.12)$$

which is graphed in Figure 3.2. This graph looks a lot like a normal distribution. In fact, I doubt that anyone could distinguish a logistic distribution from a normal distribution with the same mean and variance just by examining plots of their density functions.

Figure 3.2 Density Function for Logistic Distribution

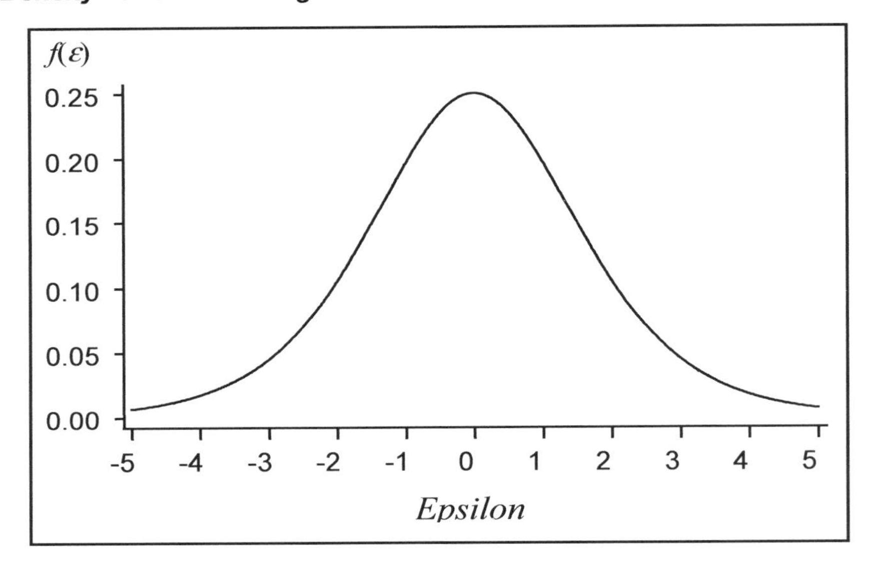

If ε has a standard logistic distribution, it follows that the dependence of y on the x's is given by the logit model:

$$\log\left[\frac{\Pr(y=1)}{\Pr(y=0)}\right] = \beta_0 + \beta_1 x_1 + \beta_2 x_2 + \ldots + \beta_k x_k \ . \qquad (3.13)$$

Furthermore, there is a simple relationship between the α's in equation (3.11) and the β's in equation (3.13):

$$\beta_0 = \frac{\alpha_0 - \mu}{\sigma}$$
$$\beta_j = \frac{\alpha_j}{\sigma}, \qquad j = 1,\ldots,k. \tag{3.14}$$

Because we don't know what σ is and have no way of estimating it, we can't directly estimate the α's. But since β_j=0 implies α_j=0 (except for the intercept), testing whether the β coefficients are 0 is equivalent to testing whether the α coefficients are 0. So the usual chi-square tests apply equally to the manifest (3.13) and latent versions (3.11) of the model.

That brings us to standardized coefficients. These can be requested with the STB option on the MODEL statement of PROC LOGISTIC. Standardized coefficients are designed to measure the relative importance of the explanatory variables in a regression model. In both linear and logistic regression, ordinary coefficients cannot generally be compared across different variables because the coefficients depend directly on the metrics in which the variables are measured. If x is measured in years, for example, its coefficient will be 12 times as large as if it were measured in months.

In linear models, one solution to this comparability problem is to convert all coefficients to standard deviation units. These standardized coefficients tell you how many standard deviations the dependent variable y changes for a 1-standard deviation increase in each of the x variables. The original coefficients are transformed to standardized coefficients by the simple formula

$$\beta_j^* = \frac{\beta_j \sigma_j}{\sigma_d}, \qquad j = 1,\ldots,k, \tag{3.15}$$

where β_j* is a standardized coefficient, σ_d is the standard deviation of the *dependent* variable, and σ_j is the standard deviation of x_j.

Will this formula work for the logit model? There's no problem multiplying each coefficient by the standard deviation of the corresponding x variable. But what should we use for σ_d? Because the observed y is dichotomous, it doesn't make a lot of sense to talk about changes in standard deviation units for this variable. Actually, the choice isn't crucial because every coefficient is divided by the same σ_d. That implies that the relative magnitudes of the standardized coefficients are the same for any choice of denominator. The choice in LOGISTIC is to take σ_d to be the standard deviation of the disturbance term ε in the latent model for the continuous variable z (equation 3.11). If ε has a standard logistic distribution, its standard deviation is $\pi/\sqrt{3} = 1.8138$. This is the value used for σ_d in equation (3.15). An implicit assumption is that the scale factor σ in equation (3.11) is equal to 1.

Output 3.28 shows the results from using the STB option with a model for the death penalty data. Let's work through the calculation of the standardized coefficient for CULP in case you ever want to do it for coefficients estimated in some other procedure (like GENMOD). The standard deviation for CULP is 1.541 (obtained with PROC MEANS). We multiply this by its coefficient (1.2709) and divide by 1.8138 to get the standardized coefficient 1.0799. As in ordinary linear regression, standardized coefficients *can* be greater than 1 or less than –1. Clearly CULP has by far the strongest effect on the outcome, followed by BLACKD and WHITVIC. Notice that this is the same rank order as the chi-square statistics. That's usually the case, but not always.

Output 3.28 LOGISTIC Output with Standardized Coefficients

Analysis of Maximum Likelihood Estimates						
Parameter	DF	Estimate	Standard Error	Wald Chi-Square	Pr > ChiSq	Standardized Estimate
Intercept	1	-5.2179	0.9298	31.4898	<.0001	
culp	1	1.2708	0.1962	41.9410	<.0001	1.0799
blackd	1	1.6358	0.5954	7.5474	0.0060	0.4525
whitvic	1	0.8476	0.5563	2.3212	0.1276	0.2298

While the standardized coefficients produced by LOGISTIC may be helpful in evaluating the relative importance of the explanatory variables, caution should be exercised when interpreting the magnitudes of these numbers. As we've just seen, the coefficients depend on a somewhat arbitrary choice for the denominator in equation (3.15). In my experience, they tend to be quite a bit larger than standardized coefficients in conventional linear models. They are also rather idiosyncratic to PROC LOGISTIC. If you include them in your research reports, don't expect others to understand them without explanation.

In addition, well-known cautions about standardized coefficients in *linear* models apply here as well. Because standardized coefficients depend heavily on the degree of variability in the explanatory variables, they can vary substantially from one population to another even when the underlying coefficients are the same. For that reason, it's dangerous to compare standardized coefficients across different groups or subgroups (Kim and Feree 1981).

3.11 Probit and Complementary Log-Log Models

The logistic model is not the only model appropriate for binary dependent variables. PROC LOGISTIC can also estimate two other widely used models, the probit model and the complementary log-log model. In this section, I'll briefly discuss each of these models, with an emphasis on how they compare with the logistic model.

The latent variable model we examined in the last section should help to motivate these models. The equation for the continuous latent variable z was

$$z = \alpha_0 + \alpha_1 x_1 + \alpha_2 x_2 + \ldots + \alpha_k x_k + \sigma\varepsilon$$

The observed variable y was equal to 1 or 0 depending on whether z was above or below some threshold value μ. We saw that if ε has a standard logistic distribution, the dependence of y on the x's is given by a logit model. Now suppose that ε has a standard *normal* distribution. That implies that y depends on the x's by way of a probit model. And if ε has a standard *extreme-value distribution* (also known as a Gumbel or double-exponential distribution), we get the complementary log-log model.

Now for the details. Both models can be seen as ways of transforming a probability into something that has no upper or lower bound. The probit model is often written as

$$\Phi^{-1}(p_i) = \beta_0 + \beta_1 x_{i1} + \beta_2 x_{i2} + \ldots + \beta_k x_{ik} \,, \qquad (3.16)$$

where p_i is the probability that y_i=1 and $\Phi^{-1}(p_i)$ is the inverse of the cumulative distribution function of a standard normal variable. This function is called the *probit function*.

Obviously, the probit function needs a little explanation. Recall that a standard normal variable has a normal distribution with a mean of 0 and a standard deviation of 1. Like every random variable, a standard normal variable has a cumulative distribution function (c.d.f.). For every possible value of the variable, the c.d.f. gives the probability that the variable is less than that value. For example, the probability that a standard normal variable is less than 1.96 is .975. You've certainly encountered the standard normal c.d.f. before—the "normal table" in the back of every introductory statistics text is either the standard normal c.d.f. or some simple variation on that function.

The *inverse* of the c.d.f. simply reverses this function. For a given probability of .90, the inverse of the c.d.f. tells you which standard normal value corresponds to a probability of .90, such that the probability of getting a number below that value is .90. In this case, the value is 1.65. We say, then, that 1.65 is the probit transformation of .90. Table 3.2 shows probit values for selected probabilities. As probabilities range between 0 and 1, the probit function ranges between $-\infty$ and $+\infty$. For comparison, I've also included the logit function and complementary log-log function (to be discussed shortly).

Table 3.2 Probit, Logit, and Complementary Log-Log Functions

Probability	Probit	Logit	C-log-log
.1	–1.28	–2.20	–2.25
.2	–.84	–1.39	–1.49
.3	–.52	–.85	–1.03
.4	–.25	–.41	–.67
.5	.00	.00	–.37
.6	.25	.41	–.09
.7	.52	.85	.19
.8	.84	1.39	.48
.9	1.28	2.20	.83

Like the logit model, the probit model is usually estimated by maximum likelihood. In PROC LOGISTIC, the way to specify a probit model is to put LINK=PROBIT as an option in the MODEL statement. (There's also a LINK=LOGIT option, but that's the default for binary data.)

Here's the LOGISTIC code for a probit model corresponding to the logit model in Output 3.28:

```
PROC LOGISTIC DATA=penalty;
  MODEL death(EVENT='1') = culp blackd whitvic / LINK=PROBIT
    STB;
RUN;
```

The results are shown in Output 3.29. The first thing to notice is that the last three columns of numbers are very similar to those shown in Output 3.28 for the corresponding logit model. In general, chi-squares, *p*-values, and standardized coefficients will usually be quite close for logit and probit models, although never identical. On the other hand, parameter estimates and standard errors are quite a bit lower for the probit model than for the logit model. Typically, logit coefficients are about 80% larger than probit coefficients. This difference has no substantive meaning and merely reflects the wider range covered by the logit transformation for a given change in the probabilities as shown in Table 3.2.

Output 3.29 LOGISTIC Output for a Probit Model

Analysis of Maximum Likelihood Estimates						
Parameter	**DF**	**Estimate**	**Standard Error**	**Wald Chi-Square**	**Pr > ChiSq**	**Standardized Estimate**
Intercept	1	-2.9932	0.4745	39.7904	<.0001	
culp	1	0.7357	0.1011	52.9402	<.0001	1.1340
blackd	1	0.9179	0.3200	8.2264	0.0041	0.4605
whitvic	1	0.4483	0.3061	2.1450	0.1430	0.2205

The standardized estimates are again calculated from the formula

$$\beta_j^* = \beta_j\left(\frac{\sigma_j}{\sigma_d}\right) \qquad j = 1,\ldots,k,$$

but now σ_d=1, the standard deviation of a standard normal distribution. So the standardized estimates are just the original coefficients multiplied by the standard deviations of the independent variables.

Is there any reason to choose a probit model over a logit model? Not really, as long as the goal is to estimate a model for a single binary dependent variable. The two models are very close and rarely lead to different qualitative conclusions. It would take very large samples to reliably discriminate between them. The logit model does have the advantage that its coefficients, when exponentiated, can be interpreted as odds ratios, something you can't do with probit coefficients. On the other hand, the probit model is more easily generalized to applications that have multiple dichotomous outcome variables. Good examples include factor analysis with dichotomous observed variables (Muthén 1984) or simultaneous equation models for dichotomous endogenous variables (Heckman 1978). Probit is attractive in these settings because the models can be based on the multivariate normal distribution, for which there is a well-developed theory and efficient computational algorithms. Multivariate logistic distributions have received much less attention.

The third model for binary data is the complementary log-log. Although not so widely known as the other two, the complementary log-log model has an important application in the area of survival analysis. The model says that

$$\log[-\log(1-p_i)] = \beta_0 + \beta_1 x_{i1} + \beta_2 x_{i2} + \ldots + \beta_k x_{ik} \qquad (3.17)$$

The expression on the left-hand side is called the complementary log-log transformation. Like the logit and the probit transformation, the complementary log-log transformation takes a number restricted to the (0, 1) interval and converts it into something with no upper or lower bound. Note that the log of $1-p$ is always a negative number. This is changed to a positive number before taking the log a second time. Solving for p, we can also write the model as

$$p_i = 1 - \exp\{-\exp[\beta_0 + \beta_1 x_{i1} + \beta_2 x_{i2} + \ldots + \beta_k x_{ik}]\}\ . \tag{3.18}$$

The complementary log-log model has one major difference from the logit and probit models. As shown in Table 3.2, while the logit and probit transformations are symmetrical around p=.50, the complementary log-log transformation is *asymmetrical*. This is also apparent from Figure 3.3 which graphs the function in equation (3.18) for a single x variable with β_0=0 and β_1=1. It's an S-shaped curve all right, but it approaches 1 much more rapidly than it approaches 0.

Figure 3.3 Plot of Complementary Log-Log Model

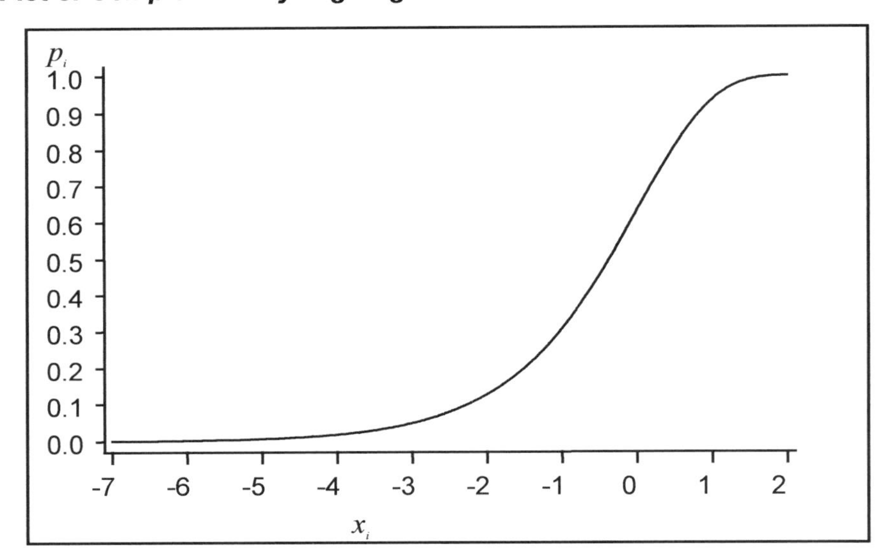

Why would you ever want an asymmetrical model? Because the complementary log-log model is closely related to continuous-time models for the occurrence of events. The most popular method of survival analysis is the *Cox regression* method, which is based on a model called the *proportional hazards model* (Cox 1972). Special cases of this model include the Weibull regression model and the exponential regression model. The proportional hazards model can be written as

$$\log h(t) = \alpha(t) + \beta_1 x_{i1} + \beta_2 x_{i2} + \ldots + \beta_k x_{ik} \tag{3.19}$$

where *h(t)* is something called the *hazard* of an event at time t and $\alpha(t)$ is an unspecified function of time. While this is not the place to discuss hazards in any detail, you can think of the hazard as the instantaneous propensity for the occurrence of events.

If you have data that contain the exact times at which events occur, you can estimate this model by a method known as *partial likelihood*. But suppose the only information you have about an event (say, a death) is whether or not it occurred in a particular year. In that case, what model should you use for the dichotomous outcome? It turns out that if events are generated by an underlying proportional hazards model (equation 3.19), then the probability of an event in some specific interval of time is given by the complementary log-log model (equation 3.18). Furthermore, the β coefficients in these models are identical. That doesn't mean you'll get the same results regardless of whether you use Cox regression for continuous data or complementary log-log analysis for discrete data. It does mean that both methods are estimating the same underlying coefficients and, therefore, have the same interpretation. For more detailed information, see Chapter 7 of my 2010 book, *Survival Analysis Using SAS: A Practical Guide, Second Edition.*

To estimate a complementary log-log model in PROC LOGISTIC, you simply put LINK=CLOGLOG as an option in the MODEL statement. Output 3.30 shows the results of doing this in LOGISTIC for a model comparable to the one in Output 3.29. With regard to qualitative conclusions, the results are quite similar to those for the logit and probit models. Moreover, all the coefficients in Output 3.30 are bracketed by the logit and the probit coefficients. In most applications I've seen, results from the complementary log-log model are not very different from logit or probit. But occasionally you'll find results that suggest qualitatively different conclusions. The standardized estimates are again calculated from equation 3.15, but now $\sigma_d = \pi/\sqrt{6} = 1.28$, the standard deviation of the standard extreme value distribution.

Output 3.30 PROC LOGISTIC Output for a Complementary Log-Log Model

Analysis of Maximum Likelihood Estimates						
Parameter	**DF**	**Estimate**	**Standard Error**	**Wald Chi-Square**	**Pr > ChiSq**	**Standardized Estimate**
Intercept	1	-4.1452	0.6713	38.1305	<.0001	
culp	1	0.8373	0.1181	50.2710	<.0001	1.0063
blackd	1	1.0975	0.4030	7.4166	0.0065	0.4293
whitvic	1	0.7040	0.3954	3.1698	0.0750	0.2700

For the death penalty data, the interpretation of the complementary log-log model in terms of an underlying proportional hazards model is really not plausible. The model isn't predicting whether an event occurs in some interval of time. Rather, given that a penalty trial takes place, it's predicting which of two kinds of outcomes occur. In cases where the model *is* plausible, we can calculate exp(β) for each coefficient to get an adjusted *hazard ratio*. For example, if we take the coefficient for BLACKD and exponentiate it, we get exp(1.0975)= 3.00. We could then say that the estimated hazard of a death sentence for a black defendant was three times the hazard for a white defendant.

Unfortunately, PROC LOGISTIC does not report the exponentiated coefficients for the complementary log-log model unless you specifically request them. And to do that, you need a separate statement for each predictor variable. For example, you can get the exponentiated coefficient (with confidence limits) for CULP with the statement:

```
ESTIMATE 'culp' culp 1 / EXP CL;
```

One final point about the complementary log-log model: because logit and probit models are symmetrical, reversing the coding of the dependent variable (from 0,1 to 1,0) only changes the signs of the coefficients. For the complementary log-log model, on the other hand, reversing the coding can give you completely different results. It's crucial, then, to set up the model to predict the probability of an *event*, not the absence of the event. If the distinction between an event and a non-event isn't obvious, then the complementary log-log model is probably not appropriate anyway.

3.12 Unobserved Heterogeneity

In Section 3.10, I explained how the logit model could be derived from a dichotomized linear model with a disturbance term that has a logistic distribution. There we saw that the logit coefficients were related to the coefficients in the underlying linear model by the formula $\beta_j = \alpha_j / \sigma$, where β_j is the logit coefficient for x_j, α_j is the corresponding coefficient in the linear model, and σ is the coefficient of the disturbance term ε. This random disturbance term can be seen as representing all omitted explanatory variables that are independent of the measured x variables, commonly referred to as *unobserved heterogeneity*. Because σ controls the variance of the disturbance, we conclude that greater unobserved heterogeneity leads to logit coefficients that are attenuated toward 0. I'll refer to this attenuation as *heterogeneity shrinkage*.

This assessment doesn't depend on the plausibility of the latent variable model. Suppose we take an ordinary logit model and put a disturbance term in the equation:

$$\log\left[\frac{\Pr(y=1\mid\varepsilon)}{\Pr(y=0\mid\varepsilon)}\right] = \beta_0 + \beta_1 x_1 + \beta_2 x_2 + \ldots + \beta_k x_k + \sigma\varepsilon. \quad (3.20)$$

We assume that ε is independent of the x's and has a standard logistic distribution, as in equation (3.12). Notice that the probabilities on the left-hand side are expressed conditionally on ε. If we then express the model unconditionally, it can be shown (Allison 1987) that the result is closely approximated by

$$\log\left[\frac{\Pr(y=1)}{\Pr(y=0)}\right] = \beta_0^* + \beta_1^* x_1 + \beta_2^* x_2 + \ldots + \beta_k^* x_k \quad (3.21)$$

where

$$\beta_j^* = \frac{\beta_j}{\sqrt{1+\sigma^2}}.$$

If we start with a probit model (and a standard normal disturbance) instead of a logit model, this result is exact rather than approximate. Again, we see that as the disturbance variance gets larger, the logit (or probit) coefficients get smaller.

Obviously, we can never measure and include all the variables that affect the dependent variable, so there will always be some heterogeneity shrinkage. What are the practical implications? First, in interpreting the results of a logistic regression, we should keep in mind that the magnitudes of the coefficients and the corresponding odds ratios are likely to be conservative to some degree. Second, to keep such shrinkage to a minimum, it's always desirable to include important explanatory variables in the model—even if we think those variables are not correlated with the variables already in the model. This advice is especially relevant to randomized experiments where there is a tendency to ignore covariates and look only at the effect of the treatment on the response. That may be okay for linear models, but logit models yield superior estimates of the treatment effect when all relevant covariates are included.

A third problem is that differential heterogeneity can confound comparisons of logit coefficients across different groups. If, for example, you want to compare logit coefficients estimated separately for men and women, you must implicitly assume that the degree of unobserved heterogeneity is the same in both groups. Elsewhere, I have proposed a method of adjusting for differential heterogeneity (Allison 1999) using a SAS macro called GLOGIT. That macro is no longer necessary, however, because the method can now be easily implemented using PROC QLIM with its HETERO statement.

Here's the example I discussed in my 1999 paper. We will estimate logistic regressions predicting the probability of promotion to associate professor for samples of 301 male and 177 female biochemists. These scientists received their doctorates in the late 1950s and early 1960s and were assistant professors at graduate departments in U.S. universities at some time during their careers (Long, Allison, and McGinnis 1993). The units of analysis are person-years rather than persons, with 1,741 person-years for men and 1,056 person-years for women. As shown in Allison (1982), the likelihood function for this sort of data factors in such a way that the multiple observations per person are effectively independent. Hence, it is entirely appropriate to use ordinary logistic regression without any correction for dependence.

The explanatory variables used in these regressions are a greatly reduced subset of the variables considered in Long et al. (1993), and the results reported here differ somewhat from those in the original article. The dependent variable PROMO is coded 1 if a person was promoted in a given year, otherwise 0. Independent variables are

FEMALE	1 for female, 0 for male.
DUR	Number of years since job began.

SELECT	A measure of selectivity of one's undergraduate institution.
ARTS	Cumulative number of articles published at each person-year.
PRESTIGE	A measure of prestige of the employing department.

We'll begin by estimating a basic logistic regression model using the syntax for PROC QLIM that was briefly presented in Chapter 2:

```
PROC QLIM DATA=promo;
  ENDOGENOUS promo~DISCRETE(DIST=LOGISTIC);
  MODEL promo = female dur dur*dur select arts prestige;
run;
```

Notice that the model includes both duration and duration squared. Results are shown in Output 3.31. All coefficients are highly significant and in the expected direction. Higher probabilities of promotion are associated with longer duration in the job, graduating from a more selective college, and having published more articles. Lower probabilities are associated with being female and being employed by a prestigious department. To get the odds ratio for FEMALE, we calculate exp(-.353751) = .70. So we can say that, in any given year, the odds that a female will be promoted is 70% of the odds that a male will be promoted.

Output 3.31 PROC QLIM Output for Promotion Data

Parameter Estimates					
Parameter	**DF**	**Estimate**	**Standard Error**	**t Value**	**Approx Pr > |t|**
Intercept	1	-6.812655	0.529019	-12.88	<.0001
FEMALE	1	-0.353751	0.132085	-2.68	0.0074
dur	1	1.723243	0.163821	10.52	<.0001
dur*dur	1	-0.125346	0.014116	-8.88	<.0001
select	1	0.154406	0.046039	3.35	0.0008
arts	1	0.054825	0.008573	6.40	<.0001
prestige	1	-0.413615	0.088254	-4.69	<.0001

We might also hypothesize that the effect of articles is greater for men than for women. We can test this by including an interaction between ARTS and FEMALE by modifying the MODEL statement as follows:

```
MODEL promo = female dur dur*dur select arts prestige
  arts*female;
```

This produced the results in Output 3.32, which clearly show some evidence for an interaction. For males, each additional article increases the odds of promotion by 100[exp(.072172)-1] = 7.5 percent. For females, each additional article increases the odds of promotion by 100[exp(.072172-.037546)-1] = 3.5 percent. This difference suggests that men get a greater payoff from their published work than do females, a conclusion that many would find troubling.

Output 3.32 PROC QLIM Output with Interaction between FEMALE and ARTS

Parameter Estimates					
Parameter	DF	Estimate	Standard Error	t Value	Approx Pr > |t|
Intercept	1	-7.000433	0.537359	-13.03	<.0001
FEMALE	1	0.009973	0.201160	0.05	0.9605
dur	1	1.720148	0.164252	10.47	<.0001
dur*dur	1	-0.125284	0.014154	-8.85	<.0001
select	1	0.152113	0.046039	3.30	0.0010
arts	1	0.072172	0.011319	6.38	<.0001
prestige	1	-0.393508	0.088585	-4.44	<.0001
FEMALE*arts	1	-0.037546	0.015789	-2.38	0.0174

The danger with this conclusion is that it assumes that the degree of residual variation is the same for men and for women. In the case of assistant professors, there is reason to believe that women have more heterogeneous career patterns than men, especially during the period covered by the data used here. Hence, unmeasured variables affecting the chances of promotion may be more important for women than for men, leading to a larger value of σ for women than men in equation (3.20). That difference could explain why the effect of articles appears to be greater for men than for women.

QLIM can allow for differing values of σ with the HETERO statement. Here's how:

```
PROC QLIM DATA=promo;
  ENDOGENOUS promo~DISCRETE(DIST=LOGISTIC);
  MODEL promo = female dur dur*dur select arts prestige
    arts*female;
  HETERO promo~female / NOCONST;
run;
```

The HETERO statement says that the disturbance variance in PROMO is allowed to be a function of FEMALE. The NOCONST option parameterizes that relationship in a way that I think is most easily interpreted. In that case, the model can be written by putting an *i* subscript on σ in equation (3.20) and writing

$$\log \sigma_i^2 = \gamma w_i \ , \tag{3.22}$$

where w_i is the dummy variable for female. Results in Output 3.33 no longer show such strong evidence for interaction: the *p*-value for the interaction is .10. On the other hand, we also don't see much evidence for differential heterogeneity by sex: the *p*-value for the effect of FEMALE on σ is .28. Although not significantly different from 0, the magnitude of that effect may be interpreted by calculating exp(.354838) = 1.43. So we say that the variance for females is estimated to be 43 percent higher than the variance for males.

Output 3.33 PROC QLIM Output with Adjustment for Heterogeneity Differences by Sex

Parameter Estimates					
Parameter	DF	Estimate	Standard Error	t Value	Approx Pr > |t|
Intercept	1	-7.365286	0.654514	-11.25	<.0001
FEMALE	1	-0.378060	0.449957	-0.84	0.4008
dur	1	1.838257	0.202864	9.06	<.0001
dur*dur	1	-0.134283	0.017016	-7.89	<.0001
select	1	0.169966	0.051663	3.29	0.0010
arts	1	0.071982	0.011411	6.31	<.0001
prestige	1	-0.420474	0.096119	-4.37	<.0001
FEMALE*arts	1	-0.030484	0.018743	-1.63	0.1039
_H.FEMALE	1	0.354838	0.325341	1.09	0.2754

These results duplicate those I described in my 1999 article. However, PROC QLIM can go a lot further by letting the disturbance variance be a function of two or more variables, either categorical or continuous. Specifically, we can let w_i in equation (3.22) be a vector of variables and γ a vector of coefficients. Here's a model that allows the disturbance variance to be a function of DUR, ARTS, and PRESTIGE:

```
PROC QLIM DATA=promo;
  ENDOGENOUS promo~DISCRETE(DIST=LOGISTIC);
  MODEL promo = female dur dur*dur select arts prestige
     arts*female;
  HETERO promo~ dur arts prestige / NOCONST;
RUN;
```

Results in Output 3.34 show evidence for effects of all three variables on the disturbance variance. Again, evidence for the interaction of FEMALE and ARTS is weak.

Output 3.34 PROC QLIM Output with Adjustment for Heterogeneity as a Function of Continuous Variables

Parameter Estimates					
Parameter	DF	Estimate	Standard Error	t Value	Approx Pr > |t|
Intercept	1	-15.783409	3.433278	-4.60	<.0001
FEMALE	1	-0.406549	0.954972	-0.43	0.6703
dur	1	5.713137	1.773517	3.22	0.0013
dur*dur	1	-0.459553	0.159871	-2.87	0.0040
select	1	0.542043	0.266106	2.04	0.0417
arts	1	0.229155	0.097094	2.36	0.0183
prestige	1	-4.488757	2.240666	-2.00	0.0451
FEMALE*arts	1	-0.115662	0.127070	-0.91	0.3627
_H.dur	1	0.172493	0.081848	2.11	0.0351
_H.arts	1	0.066629	0.023273	2.86	0.0042
_H.prestige	1	0.625576	0.190895	3.28	0.0010

Heterogeneity shrinkage is characteristic of a wide variety of non-linear regression models (Gail et al. 1984). The phenomenon is closely related to a distinction that is commonly made between *population-averaged models* and *subject-specific* models. The model that explicitly includes the disturbance term equation (3.20) is called a subject-specific model, because its coefficients describe how the log-odds changes if the explanatory variables are changed for that particular individual. On the other hand, equation (3.21)—which integrates out the disturbance term—is called a population-averaged model because its coefficients describe what happens to a whole population when the explanatory variables are changed for all individuals. If our interest is in understanding the underlying causal mechanism, then subject-specific coefficients are of primary interest. If the aim is to determine the aggregate consequences of some policy change that affects everyone, then population-averaged coefficients may be more appropriate. Keep in mind, however, that there is only one true subject-specific model, but the population-averaged model may change with the population and the degree of unobserved heterogeneity.

Can we ever estimate the subject-specific model or can we only approximate it by including more and more explanatory variables? If we have only one observation per individual, the subject-specific model cannot be directly estimated. But if there are two or more observations per individual, the parameter σ is identified, which means that the subject-specific coefficients can be recovered. We'll discuss methods for accomplishing that in Chapter 8. Unobserved heterogeneity is also closely related to a phenomenon known as *overdispersion*, which arises when estimating a logit model using grouped data. We'll discuss overdispersion in Chapter 4.

3.13 Sampling on the Dependent Variable

The logit model has a unique sampling property that is extremely useful in a variety of situations. In the estimation of linear models, sampling on the dependent variable is widely known to be potentially dangerous. In fact, much of the literature on selection bias in linear models has to do with fixing the problems that arise from such sampling (Heckman 1979). That's not true for the logit model, however. You can do disproportionate stratified random sampling on the dependent variable without biasing the coefficient estimates.

Here's a simple example. Table 3.3 shows a hypothetical table for employment status by high school graduation. The odds ratio for this table is $(570\times52)/(360\times22)=3.74$. If we estimated a logit model predicting employment status from high school graduation, the coefficient would be $\log(3.74)=1.32$. (If we reversed the independent and dependent variables, the logit coefficient would still be 1.32.) Now suppose we take a 10% random sample from the employed column and delete the other 90% of employed persons from the table. Ignoring sampling variability, the numbers in the employed column would change to 57 and 36. The new odds ratio is $(57\times52)/(36\times22)=3.74$ and, of course, the logit coefficient is still 1.32. We see, then, that sampling on the dependent variable doesn't change odds ratios.

Table 3.3 Employment Status by High School Graduation

High School Graduate	Employment Status	
	Employed	Unemployed
Yes	570	22
No	360	52

This property of contingency tables has been known for decades. Eventually, it was extended to logistic regression models with continuous independent variables (Prentice and Pyke 1979); the slope coefficients are not biased by disproportionate random sampling on the dependent variable. The intercept does change under such sampling schemes, but ordinarily we don't care much about the intercept. And, as we'll see below, there's a simple formula for getting an approximately unbiased estimate for the intercept as well.

This property of logistic regression has a couple of important practical implications. Suppose you have a census data set with 10 million observations. You want to analyze the determinants of employment status, but you don't want to stare at your computer screen while LOGISTIC chugs through 10 million cases. So you take a 0.1% simple random sample, giving you 10,000 cases. If the unemployment rate is 5%, you would expect to get about 9500 employed cases and 500 unemployed cases. Not bad, but you can do better. For a given sample size, the standard errors of the coefficients depend heavily on the split on the dependent variable. As a general rule, you're much better off with a 50-50 split than with a 95-5 split. The solution is to take a 10% sample of the unemployed cases and a 0.5% sample of the employed cases. That way you end up with about 5,000 cases of each, which will give you much smaller standard errors for the coefficients.

After you've estimated the logit model from the disproportionately stratified sample, you can easily correct the intercept to reflect the true population proportions (Hosmer and Lemeshow 2000). For example, suppose you estimated a model predicting the probability of unemployment and you got an intercept of b_0. Suppose, further, that the sampling fraction for unemployed persons was p_u, and the sampling fraction for employed persons was p_e. Then, the corrected intercept is b_o–log(p_u/p_e).

The other application of sampling on the dependent variable is the case-control study, widely used in the biomedical sciences (Breslow and Day 1980) but occasionally found in the social sciences as well. Here's how it works. You want to study the determinants of some rare disease. By surveying doctors and hospitals, you manage to identify all 75 persons diagnosed with that disease (the cases) in your metropolitan area. To get a comparison group, you recruit a 1 percent simple random sample of healthy persons in the general population of the metropolitan area (the controls). You pool the cases and the controls into a single sample and do a logistic regression predicting case vs. control, based on various background characteristics. If done correctly, this should yield approximately unbiased estimates of the population logistic regression coefficients. Often the controls are matched to the cases on one or more criteria, but that's not an essential feature of the design. In Chapter 8, we'll see how to analyze matched case-control data.

3.14 Plotting Effects of Predictor Variables

Because of the inherent nonlinearity of the logistic regression model, it's often difficult to visualize the effect of a predictor variable on the probability of an event. The EFFECTPLOT statement provides an elegant solution to that problem. It's particularly useful in examining interaction effects and polynomial functions.

I will use the academic promotion data of Section 3.12 to illustrate some of the many graphs that can be produced by the EFFECTPLOT statement. Let's begin with a model that contains no interactions or polynomial functions:

```
PROC LOGISTIC DATA=promo;
  MODEL promo(EVENT='1') = female dur select arts;
  EFFECTPLOT FIT(X=arts);
run;
```

The EFFECTPLOT statement requests a plot of the predicted probabilities of promotion as a function of ARTS (number of published articles), while setting all the other variables at their mean values. If there were any CLASS variables in the model, they would be set to their reference levels.

The graph is shown in Output 3.35. The horizontal axis includes the full range of observed values of ARTS. If you put the option / LINK on the EFFECTPLOT statement, you get a graph with the linear predictor (the logarithm of the odds) on the vertical axis which, in this case, would be a straight line. The small circles at 0 and 1 on the horizontal axis represent the actual values of the dependent variable at different values of the predictor. In this case, a single circle may represent many data points.

Output 3.35 Predicted Probability of Promotion as a Function of Number of Articles

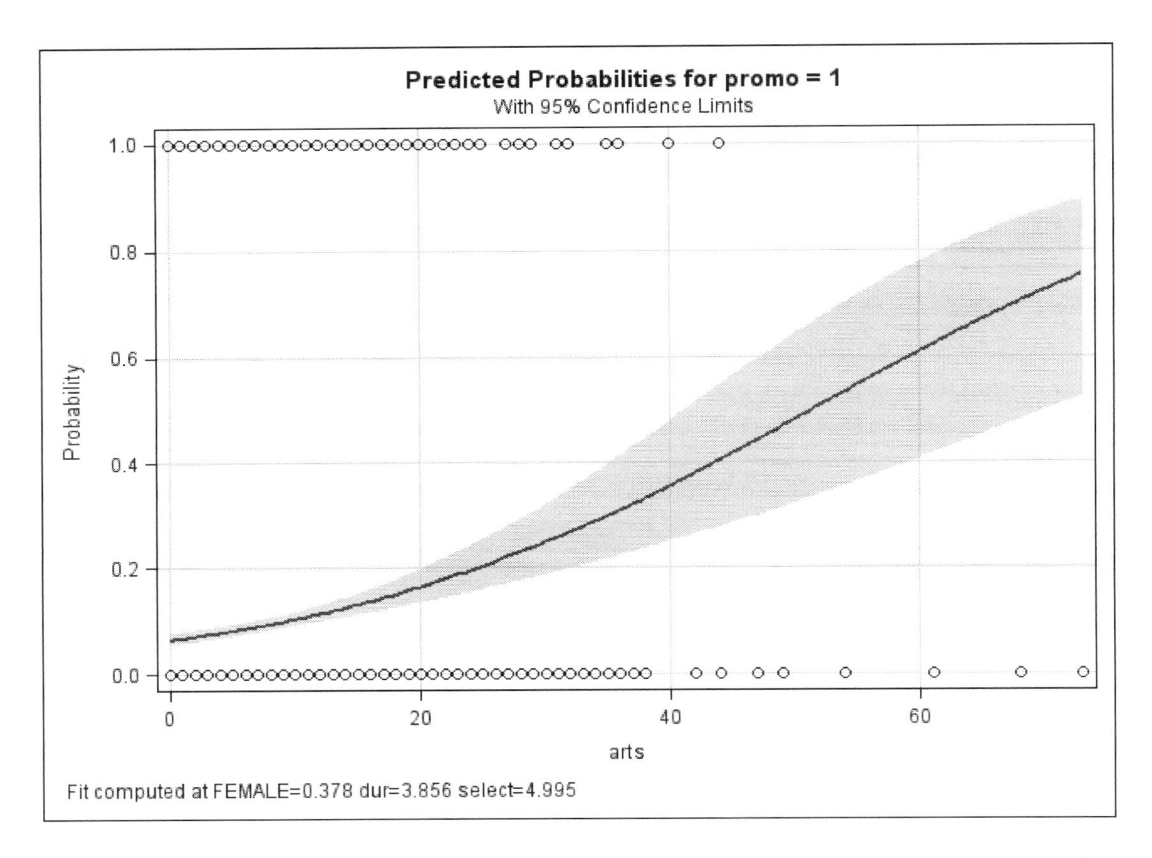

Somewhat surprisingly, the graph shows that no promotions occurred for any of the observations with more than 44 articles, despite the fact that the predicted probability of a promotion is an increasing function of number of articles. This suggests that the model may be misspecified, and that we ought to allow for a reversal of direction in the effect of ARTS. We can do this by adding the square of ARTS to the model:

```
PROC LOGISTIC DATA=promo;
  MODEL promo(EVENT='1') = female dur select arts arts*arts;
  EFFECTPLOT FIT(X=arts);
run;
```

The coefficient for ARTS*ARTS is highly significant ($p < .0001$), indicating a departure from linearity (on the log-odds scale). The graph in Output 3.36 shows that the probability of a promotion rises to a maximum at about 26 articles and then declines. Keep in mind that 26 articles is the 98th percentile, so caution is warranted before concluding that there is a real decline at these very high levels of publication.

Output 3.36 Predicted Probability of Promotion as a Function of Number of Articles

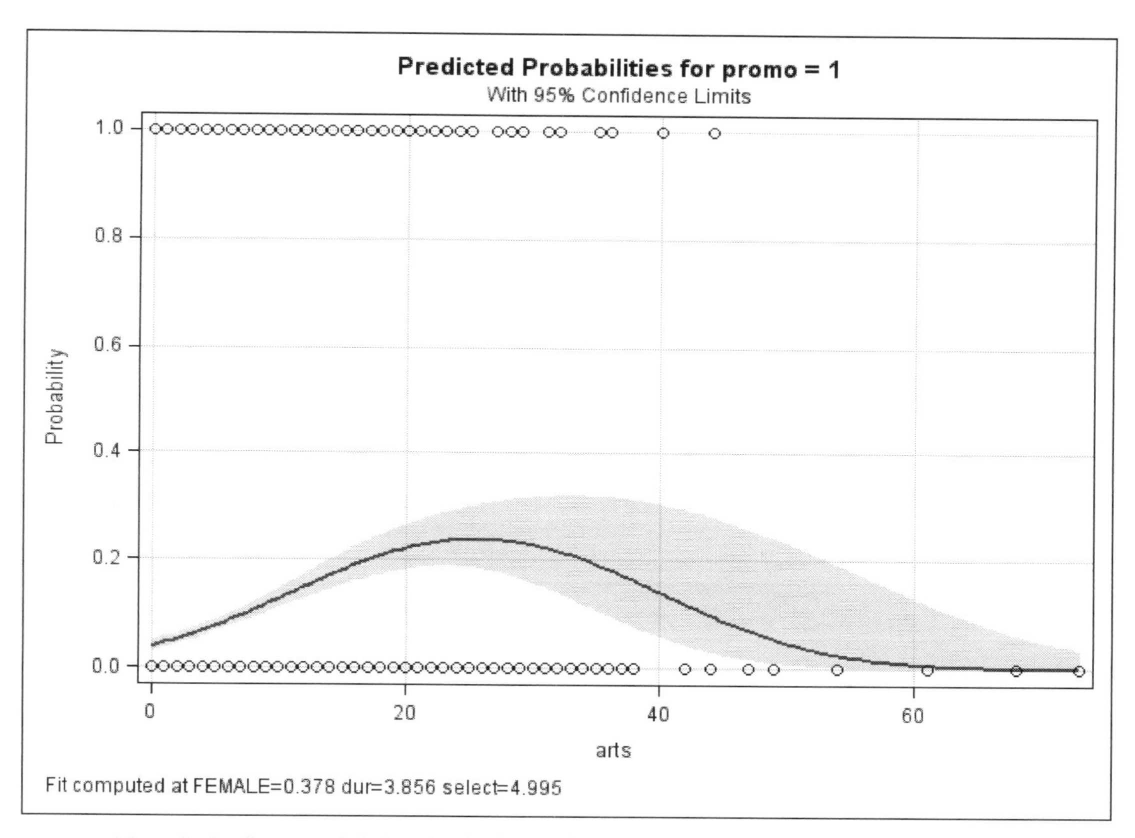

Now let's fit a model that includes an interaction between FEMALE and ARTS. To keep things simple, I'll leave out the squared term for ARTS. Here we'll use the SLICEFIT option to produce a graph that displays separate curves for men and women:

```
PROC LOGISTIC DATA=promo;
  MODEL promo(EVENT='1') = female dur select arts prestige
    arts*female ;
  EFFECTPLOT SLICEFIT(X=arts SLICEBY=female=0 1);
run;
```

The interaction is significant with a *p*-value of about .01. Output 3.37 gives us a good sense of the nature of this interaction. At low levels of article production, men and women have about equal predicted probabilities of promotion. As the number of articles increases, however, men do better than women, and the gap steadily increases. In Section 3.12, however, we saw that the evidence for this interaction becomes very weak once we adjust for differential residual heterogeneity. Furthermore, if we add the squared term for articles to the model, the interaction between FEMALE and ARTS disappears almost completely.

Output 3.37 Predicted Probability of Promotion as a Function of Articles, by Gender

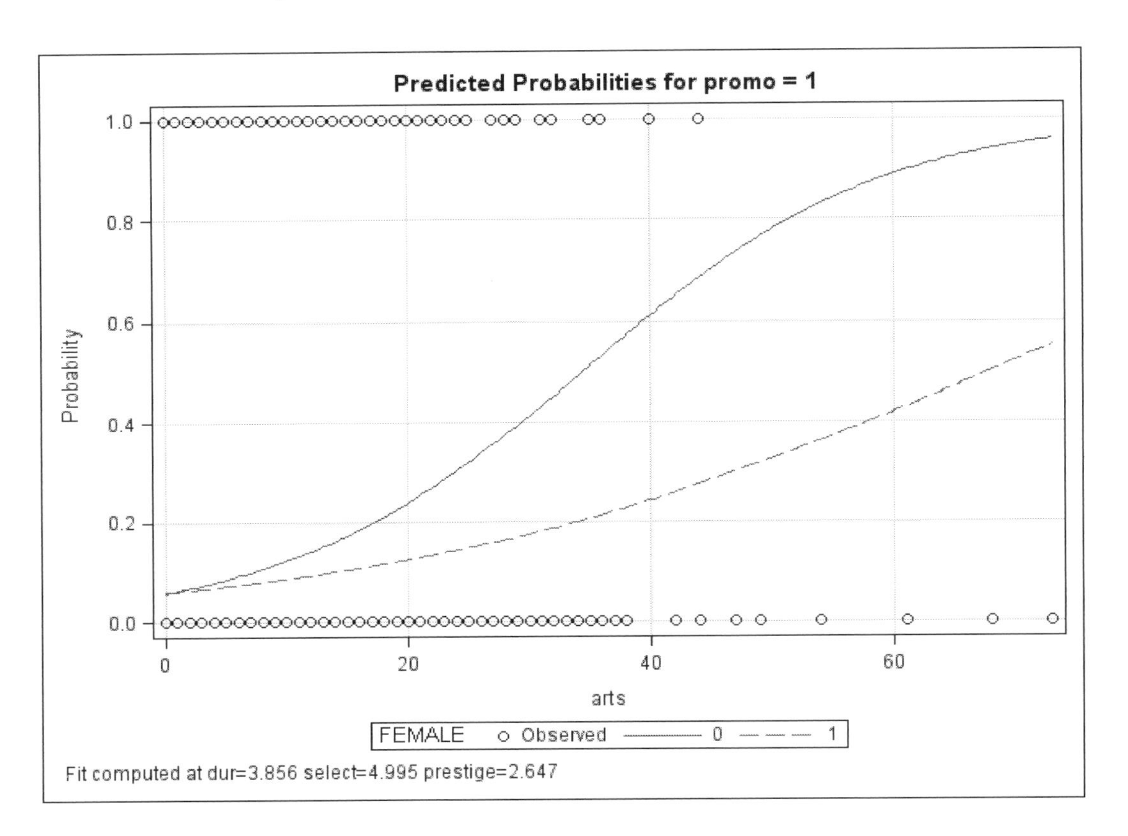

One coding note: If FEMALE had been declared as a CLASS variable, we wouldn't need the "=0 1" in the SLICEBY option. However, if a non-CLASS variable is specified as the SLICEBY variable and no numbers are given, the default is to plot the function at five equally spaced values of the SLICEBY variable. That wouldn't be sensible here because the values would be 0, .25, .50, .75, and 1, but the variable itself can only take on values of 0 and 1.

Chapter 4

Logit Analysis of Contingency Tables

4.1 Introduction

To many people, categorical data analysis means the analysis of contingency tables, also known as cross tabulations. For decades, the mainstay of contingency table analysis was the chi-square test introduced by Karl Pearson in 1900. However, things changed dramatically with the development of the loglinear model in the late 1960s and early 1970s. Loglinear analysis made it possible to analyze multi-way contingency tables, testing both simple and complex hypotheses in an elegant statistical framework. In Chapter 10, we'll see how to estimate loglinear models using the GENMOD procedure.

Although loglinear analysis is still a popular approach to the analysis of contingency tables, logit analysis (logistic regression) can often do a better job. In fact, there is an intimate relationship between the two approaches. For a contingency table, every logit model has a loglinear model that is its exact equivalent. The converse doesn't hold—there are loglinear models that don't correspond to any logit models—but in most cases, such models have little substantive interest.

If logit and loglinear models are equivalent, why use the logit model? Here are three reasons:

- The logit model makes a clear distinction between the dependent variable and the independent variables. The loglinear model makes no such distinction—all the conceptual variables are on the right-hand side of the equation.
- The loglinear model has many more parameters than the corresponding logit model. Most of these are nuisance parameters that have no substantive interest, and their inclusion in the model can be confusing (and potentially misleading).
- With larger tables, loglinear models are much more prone to convergence failures because of cell frequencies of 0.

In short, logit analysis can be much simpler than loglinear analysis even when estimating equivalent models. As we shall see, logit analysis of contingency tables has more in common with ordinary multiple regression than it does with traditional chi-square tests. In the remainder of this chapter, we'll see how to analyze contingency tables using the LOGISTIC procedure. In this chapter, we'll look only at tables where the dependent variable is dichotomous, but later chapters will consider tabular data with dependent variables having more than two categories. Much of what we do in this chapter will involve the application of tools already developed in preceding chapters.

4.2 A Logit Model for a 2 × 2 Table

Let's begin with the simplest case—a dichotomous dependent variable and a single dichotomous independent variable. That leads to a 2 × 2 table like the one in Table 2.2, reproduced here as Table 4.1.

Table 4.1 Death Sentence by Race of Defendant

	Blacks	Nonblacks	Total
Death	28	22	50
Life	45	52	97
Total	73	74	147

If we have access to the individual-level data, we can simply estimate a logit model directly, as with the following LOGISTIC program:

```
PROC LOGISTIC DATA=penalty;
  MODEL death(EVENT='1') = blackd;
RUN;
```

Results are shown in Output 4.1. The odds ratio is identical to what we got in Section 2.4 by hand calculation. It is not statistically significant.

Output 4.1 PROC LOGISTIC Output for a 2 x 2 Table

Testing Global Null Hypothesis: BETA=0			
Test	Chi-Square	DF	Pr > ChiSq
Likelihood Ratio	1.2205	1	0.2693
Score	1.2184	1	0.2697
Wald	1.2134	1	0.2707

Analysis of Maximum Likelihood Estimates					
Parameter	DF	Estimate	Standard Error	Wald Chi-Square	Pr > ChiSq
Intercept	1	-0.8602	0.2543	11.4390	0.0007
blackd	1	0.3857	0.3502	1.2134	0.2707

Odds Ratio Estimates			
Effect	Point Estimate	95% Wald Confidence Limits	
blackd	1.471	0.740	2.921

Now, suppose we don't have the individual-level data. All we have is Table 4.1. There are two different ways to get the tabular data into PROC LOGISTIC. The first method is to create a data set that has one record for each of the four cell frequencies in the table, together with appropriately coded variables that represent the defendant's race and sentence:

```
DATA tab4_1a;
  INPUT f blackd death;
  DATALINES;
22 0 1
28 1 1
52 0 0
45 1 0
;
```

Then we use the FREQ statement in LOGISTIC to specify that each of these data lines is to be replicated, using the number of replications given by the variable F:

```
PROC LOGISTIC DATA=tab4_1a;
  FREQ f;
  MODEL death(EVENT='1') = blackd
RUN;
```

Results are *identical* to those in Output 4.1.

The other way to handle tabular data is to use the *events/trials* syntax. Instead of inputting all four internal cell counts in the table, we input the cell frequencies for death sentences along with the column totals, i.e., the number of black defendants and the number of non-black defendants:

```
DATA tab4_1b;
  INPUT death total blackd;
  DATALINES;
22 74 0
28 73 1
;
```

We can describe this data set as follows: among 74 non-black defendants, 22 got the death penalty; among 73 black defendants, 28 got the death penalty.

Next we specify a dependent variable that has two parts, separated by a slash:

```
PROC LOGISTIC DATA=tab4_1b;
  MODEL death/total = blackd;
RUN;
```

Instead of replicating the observations, LOGISTIC treats the variable DEATH as having a binomial distribution with the number of trials (an apt term for these data) given by the variable TOTAL. Note that with this syntax, you do *not* need the EVENT='1' option in the MODEL statement. That's because using DEATH as the number of events directly specifies death penalty (rather than life sentence) as the event of interest. Again, results are identical to those in Output 4.1.

For a 2 × 2 table, we can also produce many of these results with PROC FREQ:

```
PROC FREQ DATA=tab4_1a;
  WEIGHT f;
  TABLES blackd*death / CHISQ RELRISK;
RUN;
```

Selected results are shown in Output 4.2.

Output 4.2 PROC FREQ Results

Statistics for Table of blackd by death

Statistic	DF	Value	Prob
Chi-Square	1	1.2184	0.2697
Likelihood Ratio Chi-Square	1	1.2205	0.2693
Continuity Adj. Chi-Square	1	0.8644	0.3525
Mantel-Haenszel Chi-Square	1	1.2101	0.2713
Phi Coefficient		0.0910	
Contingency Coefficient		0.0907	
Cramer's V		0.0910	

Estimates of the Relative Risk (Row1/Row2)			
Type of Study	Value	95% Confidence Limits	
Case-Control (Odds Ratio)	1.4707	0.7404	2.9214

The CHISQ option gives us several statistics for testing the null hypothesis that the two variables are independent, which is equivalent to the hypothesis that the slope is 0 in the logit model. Notice that Pearson's chi-square (labeled simply chi-square in Output 4.2) is identical to the score statistic in the PROC LOGISTIC Output 4.1. The likelihood ratio chi-square in Output 4.2 is identical to the chi-square labeled –2LOG L in Output 4.1. The RELRISK option gives us the odds ratio, which is identical to the odds ratio in the PROC LOGISTIC output.

4.3 A Three-Way Table

As we've just seen, using LOGISTIC to estimate a logit model for a 2 × 2 table doesn't tell you much more than you would get with PROC FREQ. But with a 3-way table, things get more interesting.

Consider Table 4.2, which shows the cross classification of race, gender, and occurrence of sexual intercourse for a national sample of 15- and 16-year-olds, reported by Morgan and Teachman (1988).

Table 4.2 Race by Gender by Sexual Intercourse

		Intercourse	
Race	Gender	Yes	No
White	Male	43	134
	Female	26	149
Black	Male	29	23
	Female	22	36

Let's analyze Table 4.2 as a logit model, with intercourse as the dependent variable and race and sex as independent variables. Using the *events/trials* syntax in LOGISTIC, we have:

```
DATA interco;
  INPUT white male yes no;
  total=yes+no;
  DATALINES;
1 1 43 134
1 0 26 149
0 1 29 23
0 0 22 36
;
PROC LOGISTIC DATA=interco;
  MODEL yes/total=white male / SCALE=NONE;
RUN;
```

The SCALE=NONE option is included so that we can get the deviance and Pearson chi-square tests for goodness of fit. When you use the *events/trials* syntax, the AGGREGATE option is not needed because the data are already in aggregated form.

Output 4.3 PROC LOGISTIC Output for Analysis of Three-Way Table

Deviance and Pearson Goodness-of-Fit Statistics				
Criterion	**Value**	**DF**	**Value/DF**	**Pr > ChiSq**
Deviance	0.0583	1	0.0583	0.8091
Pearson	0.0583	1	0.0583	0.8092

Analysis of Maximum Likelihood Estimates					
Parameter	**DF**	**Estimate**	**Standard Error**	**Wald Chi-Square**	**Pr > ChiSq**
Intercept	1	-0.4555	0.2221	4.2045	0.0403
white	1	-1.3135	0.2378	30.5132	<.0001
male	1	0.6478	0.2250	8.2872	0.0040

Odds Ratio Estimates			
Effect	**Point Estimate**	**95% Wald Confidence Limits**	
white	0.269	0.169	0.429
male	1.911	1.230	2.971

Selected results are shown in Output 4.3. The first thing to notice is that the deviance and Pearson chi-square statistics are both only .058 with 1 degree of freedom, yielding a *p*-value of .81. Because this is a goodness-of-fit test, a high *p*-value means that the model fits well. Clearly this model fits the data extremely well.

But what is the deviance testing? Remember that the deviance always compares the fitted model with a saturated model. For a $2 \times 2 \times 2$ table, the saturated logit model includes an interaction between the two explanatory variables in their effects on the dependent variable. There's one coefficient for this interaction, corresponding to the single degree of freedom for the deviance. So, in this case, the deviance is testing the null hypothesis that the interaction coefficient is 0. We conclude that there's no evidence for an interaction.

To see this more explicitly, let's actually fit the model with the interaction:

```
PROC LOGISTIC DATA=interco;
  MODEL yes/total=white male white*male/ SCALE=NONE   ;
RUN;
```

In Output 4.4 we see that the deviance is 0, as it should be for a saturated model. The interaction term has a chi-square of .0583 with a *p*-value of .8092, identical to what we got for the deviance in the "main effects" model. The absence of an interaction means that race has the same effect for both boys and girls. And gender has the same effect for both blacks and whites.

Output 4.4 PROC LOGISTIC Output for a Saturated Model

Deviance and Pearson Goodness-of-Fit Statistics				
Criterion	**Value**	**DF**	**Value/DF**	**Pr > ChiSq**
Deviance	0.0000	0	.	.
Pearson	0.0000	0	.	.

Analysis of Maximum Likelihood Estimates					
Parameter	**DF**	**Estimate**	**Standard Error**	**Wald Chi-Square**	**Pr > ChiSq**
Intercept	1	-0.4925	0.2706	3.3118	0.0688
white	1	-1.2534	0.3441	13.2674	0.0003
male	1	0.7243	0.3888	3.4696	0.0625
white*male	1	-0.1151	0.4765	0.0583	0.8092

What exactly are those effects? In Output 4.3, the odds ratio for MALE is 1.91. We can say, then, that the estimated odds of having had sexual intercourse by age 15 are nearly twice as large for males as for females, after adjusting for racial differences. It would probably be a mistake to put too much faith in the exact value 1.91, however. The 95% confidence interval is between 1.23 and 2.97.

For WHITE, the highly significant adjusted odds ratio is .269, indicating that the odds of intercourse for whites is a little more than one-fourth the odds for blacks. It may be easier to interpret the reciprocal 1/.269 = 3.72, which says that the odds for blacks are nearly four times the odds for whites, controlling for gender differences. This has a 95% confidence interval of 2.3 to 5.9 (obtained by taking the reciprocal of the confidence limits in Output 4.3). That's about all we can usefully do with this table. We could examine residuals but because the model fits so well, there's little point.

4.4 A Four-Way Table

Things are only a little more complicated with 4-way tables except that more interactions are possible. Consider Table 4.3, reported by Seeman (1977) based on surveys of adult males conducted in France and the United States. In each country, the respondent was asked about his own occupation, his father's occupation, and whether he considered himself a member of the working class. Occupations were classified into manual and non-manual. Our goal is to estimate a logit model for the dependence of working class identification on the other three variables.

Table 4.3 Identification with the Working Class by Country and Occupation

Country	Occupation	Father's Occupation	Identifies with the Working Class: Yes	Identifies with the Working Class: No	Total
France	Manual	Manual	85	22	107
		Non-manual	44	21	65
	Non-manual	Manual	24	42	66
		Non-manual	17	154	171
U.S.	Manual	Manual	24	63	87
		Non-manual	22	43	65
	Non-manual	Manual	1	84	85
		Non-manual	6	142	148

Here's a DATA step for reading in the table data:

```
DATA working;
  INPUT france manual famanual total working;
  DATALINES;
1   1   1  107  85
1   1   0   65  44
1   0   1   66  24
1   0   0  171  17
0   1   1   87  24
0   1   0   65  22
0   0   1   85   1
0   0   0  148   6
;
```

Using the *events/trials* syntax, let's first consider a model with no interactions:

```
PROC LOGISTIC DATA=working;
  MODEL working/total = france manual famanual / SCALE=NONE;
RUN;
```

I could have specified the explanatory variables as CLASS variables, but there's no advantage in doing that for dichotomous variables coded 1 and 0.

The results in Output 4.5 show highly significant effects of all three variables. Not surprisingly, higher probabilities of working class identification are found in France, among men who work in manual occupations and among men whose fathers worked in manual occupations. But the model doesn't fit the data very well. With a deviance of 18.98 on 4 degrees of freedom, the *p*-value is only .0008. Clearly something is missing from the model.

Output 4.5 PROC LOGISTIC Output for Working Class Identification

Deviance and Pearson Goodness-of-Fit Statistics				
Criterion	**Value**	**DF**	**Value/DF**	**Pr > ChiSq**
Deviance	18.9759	4	4.7440	0.0008
Pearson	18.9801	4	4.7450	0.0008

Analysis of Maximum Likelihood Estimates					
Parameter	**DF**	**Estimate**	**Standard Error**	**Wald Chi-Square**	**Pr > ChiSq**
Intercept	1	-3.6901	0.2547	209.9401	<.0001
france	1	1.9474	0.2162	81.1501	<.0001
manual	1	2.5199	0.2168	135.1487	<.0001
famanual	1	0.5522	0.2017	7.4964	0.0062

Odds Ratio Estimates			
Effect	**Point Estimate**	**95% Wald Confidence Limits**	
france	7.011	4.589	10.709
manual	12.427	8.126	19.006
famanual	1.737	1.170	2.579

What's missing are the interaction terms: three 2-way interactions and one 3-way interaction. Each has 1 degree of freedom, giving us the 4 degrees of freedom for the deviance. Because 3-way interactions are a pain to interpret, let's see if we can get by with just the 2-way interactions:

```
PROC LOGISTIC DATA=working;
  MODEL working/total = france manual famanual
    france*manual france*famanual manual*famanual /
    SCALE=NONE;
RUN;
```

Output 4.6 Model with All Two-Way Interactions

Deviance and Pearson Goodness-of-Fit Statistics				
Criterion	**Value**	**DF**	**Value/DF**	**Pr > ChiSq**
Deviance	3.1512	1	3.1512	0.0759
Pearson	2.8835	1	2.8835	0.0895

Analysis of Maximum Likelihood Estimates					
Parameter	**DF**	**Estimate**	**Standard Error**	**Wald Chi-Square**	**Pr > ChiSq**
Intercept	1	-3.5075	0.4269	67.4946	<.0001
france	1	1.4098	0.4585	9.4563	0.0021
manual	1	2.9517	0.4612	40.9590	<.0001
famanual	1	0.0879	0.4869	0.0326	0.8568
france*manual	1	-0.2311	0.4966	0.2166	0.6416
france*famanual	1	1.3375	0.4364	9.3945	0.0022
manual*famanual	1	-0.5968	0.4370	1.8655	0.1720

In Output 4.6 we see that the model with all three 2-way interactions fits reasonably well. The *p*-value for the deviance chi-square of 3.15 with 1 degree of freedom is .08—not great but still acceptable. The single degree of freedom corresponds to the excluded 3-way interaction. In essence, the deviance is testing whether or not this interaction is 0. Examining the Wald chi-squares for the three 2-way interactions, we find that country × father's occupation is highly significant but the other two are far from significant. That suggests fitting a model with just one 2-way interaction, for which results are shown in Output 4.7.

Output 4.7 Model with One Two-Way Interaction

Deviance and Pearson Goodness-of-Fit Statistics				
Criterion	**Value**	**DF**	**Value/DF**	**Pr > ChiSq**
Deviance	5.3744	3	1.7915	0.1463
Pearson	5.2649	3	1.7550	0.1534

Analysis of Maximum Likelihood Estimates					
Parameter	DF	Estimate	Standard Error	Wald Chi-Square	Pr > ChiSq
Intercept	1	-3.1796	0.2713	137.3766	<.0001
france	1	1.1772	0.2876	16.7546	<.0001
manual	1	2.5155	0.2162	135.3425	<.0001
famanual	1	-0.3802	0.3211	1.4021	0.2364
france*famanual	1	1.5061	0.4098	13.5068	0.0002

Odds Ratio Estimates			
Effect	Point Estimate	95% Wald Confidence Limits	
manual	12.373	8.099	18.903

When the two nonsignificant interactions are deleted, the deviance goes up but the degrees of freedom goes up even more, giving us an improved *p*-value of about .15. Moreover, the difference in deviance between the models in Output 4.6 and Output 4.7 is only 2.22 with 2 degrees of freedom, far from significant. So deleting the two interactions does not significantly worsen the fit.

Let's interpret the estimates in Output 4.7. The estimate for MANUAL is the most straightforward because it's not involved in the interaction. With an odd ratio of 12.373, we can say that manual workers have an odds of identification with the working class that is more than 12 times the odds for non-manual workers.

No odds ratios are reported, however, for the two variables with an interaction effect, FRANCE and FAMANUAL. By default, PROC LOGISTIC does not report odds ratios for variables that are involved in interactions. In the previous edition of this book, I showed how to get the desired odds ratios by doing some tedious hand calculations. But now it's much easier to let SAS do it, using the ODDSRATIO statement.

As with any 2-way interaction, we can view the results in two different ways. The model says that the effect of FAMANUAL depends on the value of FRANCE, but it also says that the effect of FRANCE depends on the value of FAMANUAL. Both views can be examined by including the following two statements in our LOGISTIC program:

```
ODDSRATIO famanual / AT(france=0 1);
ODDSRATIO france / AT(famanual=0 1);
```

which produces Output 4.8.

Output 4.8 Odds Ratios for Variables in a Two-Way Interaction

Odds Ratio Estimates and Wald Confidence Intervals			
Label	**Estimate**	**95% Confidence Limits**	
famanual at france=0	0.684	0.364	1.283
famanual at france=1	3.083	1.866	5.095
france at famanual=0	3.245	1.847	5.702
france at famanual=1	14.633	8.020	26.697

In the U.S. (FRANCE=0), the odds ratio for FAMANUAL is .684, meaning that men whose fathers had manual occupations had a 32 percent lower odds of identifying with the working class. But this odds ratio is not significantly different from 1, as indicated by the 95 percent confidence interval that easily includes 1. On the other hand, for men in France the odds ratio for FAMANUAL is 3.083 with a 95 percent confidence interval that clearly does not include 1. So in France, men whose fathers held manual occupations had an odds of working class identification that was about triple the odds of men whose fathers held non-manual jobs.

While this is the most natural way of interpreting the interaction, there's also a different view. The next two lines of Output 4.8 tell us how the effect of being in France rather than the U.S. varies by father's occupation. For men whose fathers had non-manual jobs, the odds of identifying with the working class among Frenchmen was 3.245 times the odds among Americans. Judging from the confidence interval, this is a highly significant effect. But the country effect is much larger among men whose fathers had manual jobs: the odds ratio increases to 14.633, which is an effect that is more than four times larger. We know that these two odds ratios are significantly different because the interaction is significant.

To sum up our analysis, manual workers and Frenchmen are more likely to identify with the working class. In France, men whose fathers had a manual occupation are also more likely to identify with the working class, an effect that is independent of one's own occupation. In the U.S., however, a father's occupation makes little, if any, difference.

4.5 A Four-Way Table with Ordinal Explanatory Variables

Next we consider a $2 \times 2 \times 4 \times 4$ table, reported by Sewell and Shah (1968), for a sample of 4,991 high school seniors in Wisconsin. The dependent variable was whether or not they planned to attend college in the following year. The three independent variables were coded as follows:

IQ	1=low, 2=lower middle, 3=upper middle, 4=high
SES	1=low, 2=lower middle, 3=upper middle, 4=high
PARENT	1=low parental encouragement, 2=high encouragement

Following is a DATA step to read in the table, in a form suitable for the *events/trials* syntax. There is one record for each combination of the independent variables. The fourth variable TOTAL is the number of students who had each set of values for the independent variables, and the last variable COLL is the number of students who planned to attend college. Thus, on the first line, we see that there were 353 students who were low on all three independent variables; of those, only four planned to attend college.

```
DATA wisc;
  INPUT iq parent ses total coll;
  DATALINES;
1   1   1   353     4
1   1   2   234     2
1   1   3   174     8
1   1   4    52     4
1   2   1    77    13
1   2   2   111    27
1   2   3   138    47
1   2   4    96    39
2   1   1   216     9
2   1   2   208     7
2   1   3   126     6
2   1   4    52     5
2   2   1   105    33
2   2   2   159    64
2   2   3   184    74
2   2   4   213   123
3   1   1   138    12
3   1   2   127    12
3   1   3   109    17
3   1   4    50     9
3   2   1    92    38
```

```
3  2  2  185   93
3  2  3  248  148
3  2  4  289  224
4  1  1   77   10
4  1  2   96   17
4  1  3   48    6
4  1  4   25    8
4  2  1   92   49
4  2  2  178  119
4  2  3  271  198
4  2  4  468  414
;
```

Now we're ready for PROC LOGISTIC:

```
PROC LOGISTIC DATA=wisc;
  CLASS iq ses / PARAM=REF;
  MODEL coll/total=iq ses parent / SCALE=NONE;
RUN;
```

IQ and SES are listed as CLASS variables so that they will be treated as categorical rather than quantitative variables. As explained in Chapter 2, the PARAM=REF option overrides the default method for coding CLASS variables. Specifically, the REF option tells LOGISTIC to create a set of dummy (indicator) variables, with the reference category being the highest value of the CLASS variable. Output 4.9 displays selected portions from the LOGISTIC output.

Output 4.9 PROC LOGISTIC Output for a Four-Way Table

Deviance and Pearson Goodness-of-Fit Statistics				
Criterion	**Value**	**DF**	**Value/DF**	**Pr > ChiSq**
Deviance	25.2358	24	1.0515	0.3930
Pearson	24.4398	24	1.0183	0.4367

Type 3 Analysis of Effects			
Effect	**DF**	**Wald Chi-Square**	**Pr > ChiSq**
iq	3	334.4508	<.0001
ses	3	174.4307	<.0001
parent	1	586.3859	<.0001

Analysis of Maximum Likelihood Estimates						
Parameter		DF	Estimate	Standard Error	Wald Chi-Square	Pr > ChiSq
Intercept		1	-3.1005	0.2123	213.3353	<.0001
iq	1	1	-1.9663	0.1210	264.2400	<.0001
iq	2	1	-1.3722	0.1024	179.7284	<.0001
iq	3	1	-0.6331	0.0976	42.0830	<.0001
ses	1	1	-1.4140	0.1210	136.6657	<.0001
ses	2	1	-1.0580	0.1029	105.7894	<.0001
ses	3	1	-0.7516	0.0976	59.3364	<.0001
parent		1	2.4554	0.1014	586.3859	<.0001

Odds Ratio Estimates			
Effect	Point Estimate	95% Wald Confidence Limits	
iq 1 vs 4	0.140	0.110	0.177
iq 2 vs 4	0.254	0.207	0.310
iq 3 vs 4	0.531	0.439	0.643
ses 1 vs 4	0.243	0.192	0.308
ses 2 vs 4	0.347	0.284	0.425
ses 3 vs 4	0.472	0.390	0.571
parent	11.651	9.551	14.213

Judging from the deviance and Pearson chi-squares, we see that a model with main effects and no interactions fits the data well, with *p*-values of .39 and .43 for the two statistics. As in the previous table, these goodness-of-fit statistics are testing the null hypothesis that all the omitted terms (three 2-way interactions and one 3-way interaction) are 0.

Despite the fact that the overall deviance is not significant, I also checked to see if any of the 2-way interactions might be significant. This could happen if most of the deviance was attributable to a single omitted interaction. The LOGISTIC code for fitting all the 2-way interactions is:

```
PROC LOGISTIC DATA=wisc;
  CLASS iq ses /PARAM=REF;
  MODEL coll/total=iq|ses|parent @2;
RUN;
```

The syntax IQ|SES|PARENT is shorthand for all possible interactions and lower-order terms among the three variables. Because I wasn't interested in the 3-way interaction, I added the @2 option, which restricts the model to 2-way interactions and main effects. In Output 4.10 we see that none of the 2-way interactions is statistically significant.

Output 4.10 Tests for Main Effects and Two-Way Interactions

Type 3 Analysis of Effects			
Effect	DF	Wald Chi-Square	Pr > ChiSq
iq	3	25.7986	<.0001
ses	3	3.6215	0.3053
iq*ses	9	12.4348	0.1899
parent	1	83.0971	<.0001
parent*iq	3	2.9139	0.4051
parent*ses	3	2.2394	0.5242

Given the large sample size, the "no interactions" model fits the data impressively well. Now we can feel confident that the main-effects model in Output 4.9 is a reasonably good fit to the data. From the Type 3 table in Output 4.9, we see that each of the three variables is highly significant, with parental encouragement having the strongest effect, followed by IQ and then SES. One nice thing about these Type 3 chi-squares is that they don't depend on the choice of the reference category. In fact, they are invariant to any choice among the several PARAM options.

Turning to the parameter estimates, the easiest one to interpret is the effect of PARENT because it only has two values. Although PARENT has values of 1 for low encouragement and 2 for high encouragement, the coefficient (and odds ratio) are identical to what they would be if the values of PARENT were 0 for low and 1 for high. The odds ratio of 11.65 says that students whose parents gave high levels of encouragement are nearly 12 times as likely to plan to attend college as students whose parents gave low levels of encouragement. Because the sample size is large, this adjusted odds ratio is estimated with good precision: the 95% confidence interval is 9.5 to 14.2.

For IQ and SES, the odds ratio estimates in Output 4.9 make it clear that each is a comparison between a particular category and the reference category (level 4 for IQ or SES). For example, the odds ratio for IQ 1 vs. 4 is .14, indicating that the odds of college plans among the low IQ group are about one seventh the odds in the high IQ group. For IQ 2 vs. 4, we have an odds ratio of .25. This tells us that students in the lower-middle IQ group have odds that are only about one-fourth the odds for the high group. Finally, comparing upper-middle and high groups, we get an odds ratio of .53. Thus, the highest group has about double the odds of college plans as the upper-middle group. From the chi-square column (in the "Analysis of Maximum Likelihood Estimates" table in Output 4.9), we see that each of these contrasts with the reference category is statistically significant at well beyond the .001 level.

For SES, each of the odds ratios is a contrast with the high SES category. Each SES level is significantly different from the highest category and, as you might expect, the odds of college plans goes up with each increase in SES. But what about comparisons between other categories? Suppose we want to know if there's a significant difference between lower-middle (SES 2) and upper-middle (SES 3) groups with respect to college plans. The magnitude of the difference is found by taking the difference in coefficients: $-.7516 - (-1.058) = .3064$. Exponentiating yields 1.36, indicating that the odds of college plans are about 36% higher in the upper-middle than in the lower-middle group. To test for significance of the difference, we could use either a CONTRAST statement or a TEST statement after the MODEL statement. I'll use CONTRAST because it can also produce estimates of the coefficient difference and the odds ratio:

```
CONTRAST 'um vs. lm' ses 0 1 -1 0 / ESTIMATE=BOTH;
```

The text within quotes is a label; a label is mandatory but it can be any text you want. The four numbers after SES tell LOGISTIC to multiply the SES 1 coefficient by 0, SES 2 by 1, SES 3 by –1, and SES 4 by 0. LOGISTIC then sums the results and does a chi-square test of the null hypothesis that the sum is equal to 0. Of course, this is equivalent to testing whether the difference between SES 2 and SES 3 is 0. The ESTIMATE=BOTH option requests both the difference between the coefficients and the corresponding odds ratio. The output clearly indicates that the difference is significant, and it also produces the estimates that we just got by hand calculation, along with 95 percent confidence intervals.

Output 4.11 Results from CONTRAST Statement

Contrast Test Results			
Contrast	**DF**	**Wald Chi-Square**	**Pr > ChiSq**
um vs. lm	1	9.1100	0.0025

Contrast Estimation and Testing Results by Row									
Contrast	**Type**	**Row**	**Estimate**	**Standard Error**	**Alpha**	**Confidence Limits**		**Wald Chi-Square**	**Pr > ChiSq**
um vs. lm	PARM	1	-0.3064	0.1015	0.05	-0.5054	-0.1074	9.1100	0.0025
um vs. lm	EXP	1	0.7361	0.0747	0.05	0.6032	0.8981	9.1100	0.0025

The pattern of coefficients for IQ and SES suggests that SES and IQ might be treated as quantitative variables, thereby obtaining a more parsimonious representation of the data. This is easily accomplished by removing those variables from the CLASS statement. I estimated two models, one removing IQ from the CLASS statement and the other removing SES from the CLASS statement. When IQ is treated as a quantitative variable, the deviance is 25.84 with 26 degrees of freedom and a *p*-value of .47. When SES is treated as quantitative, the deviance is 35.24 with 26 degrees of freedom and a *p*-value of .11. While this is still an acceptable fit, it suggests that perhaps the effect of SES is not quite linear. To get a more sensitive test, we can take the difference in deviances between the model with SES as categorical (25.24) and the model with SES as quantitative (35.24), yielding a chi-square of 10 with 2 degrees of freedom (the difference in degrees of freedom for the two models). This test has a *p*-value of .0067, telling us that the model with SES as categorical fits significantly better than the model with SES as quantitative. (Note: It's not legitimate to compare models by taking differences in the *Pearson* chi-square; you must use differences in deviance.)

Output 4.12 gives the parameter estimates and odds ratios for the model with SES as categorical and IQ as quantitative. By removing IQ from the CLASS statement, we impose the restriction that each one-level jump in IQ has the same effect on the odds of planning to go to college. The odds ratio of 1.95 tells us that each one-level increase in IQ approximately doubles the odds (controlling for SES and parental encouragement). Consequently, moving three steps (from IQ 1 to IQ 4) multiplies the odds by 1.95^3=7.4.

Output 4.12 Model with Quantitative Effect of IQ

Analysis of Maximum Likelihood Estimates						
Parameter		DF	Estimate	Standard Error	Wald Chi-Square	Pr > ChiSq
Intercept		1	-5.7635	0.2265	647.7589	<.0001
iq		1	0.6682	0.0367	332.0319	<.0001
ses	1	1	-1.4131	0.1209	136.5537	<.0001
ses	2	1	-1.0566	0.1028	105.5986	<.0001
ses	3	1	-0.7486	0.0975	58.9692	<.0001
parent		1	2.4532	0.1013	586.0945	<.0001

Odds Ratio Estimates			
Effect	Point Estimate	95% Wald Confidence Limits	
iq	1.951	1.815	2.096
ses 1 vs 4	0.243	0.192	0.308
ses 2 vs 4	0.348	0.284	0.425
ses 3 vs 4	0.473	0.391	0.573
parent	11.625	9.531	14.179

The coefficients for SES and PARENT are about the same as they were in Output 4.9. Examination of the SES coefficients yields some insight into why couldn't we impose a linear effect of SES. If the effect of SES were linear, the difference between adjacent coefficients should be the same at every level. Yet, while the difference between SES 4 and SES 3 is .75, the difference between SES 3 and SES 2 is only .31. With a sample size of nearly 5,000, even small differences like this may show up as statistically significant.

Before leaving the 4-way table, let's consider one more model. It's reasonable to argue that parental encouragement is an intervening (mediating) variable in the production of college plans. Students with high IQ and high SES are more likely to get parental encouragement for attending college. As a result, controlling for parental encouragement may obscure the overall impact of these two variables. To check this possibility, Output 4.13 displays a model that excludes PARENT. The coefficient for IQ increases slightly—on the odds scale, it's about 14% larger. For SES the change is more dramatic. The coefficient for SES 1 has a magnitude of 2.13 in Output 4.13 compared with 1.41 in Output 4.12, corresponding to odds ratios of 8.41 vs. 4.10. So, the effect of going from the lowest to highest category of SES more than doubles when PARENT is removed from the model. It appears, then, that parental encouragement mediates a substantial portion of the overall effect of SES on college plans.

Output 4.13 Model without PARENT

Analysis of Maximum Likelihood Estimates						
Parameter		DF	Estimate	Standard Error	Wald Chi-Square	Pr > ChiSq
Intercept		1	-1.5724	0.1123	196.1822	<.0001
iq		1	0.7953	0.0335	565.1184	<.0001
ses	1	1	-2.1340	0.1091	382.5562	<.0001
ses	2	1	-1.5361	0.0935	269.6875	<.0001
ses	3	1	-0.9776	0.0894	119.4662	<.0001

Odds Ratio Estimates			
Effect	Point Estimate	95% Wald Confidence Limits	
iq	2.215	2.075	2.365
ses 1 vs 4	0.118	0.096	0.147
ses 2 vs 4	0.215	0.179	0.259
ses 3 vs 4	0.376	0.316	0.448

4.6 Overdispersion

When estimating logistic models with grouped data, it often happens that the model doesn't fit—the deviance and Pearson chi-square are large, relative to the degrees of freedom. Lack of fit is sometimes described as *overdispersion*. Overdispersion has two possible causes:

- An incorrectly specified model: more interactions and/or nonlinearities are needed in the model.
- Lack of independence of the observations: this can arise from unobserved heterogeneity that operates at the level of groups rather than individuals.

We've already seen examples of the first cause. Now let's look at an example where overdispersion may arise from dependence among the observations. The sample consists of 40 U.S. biochemistry departments in the late 1950s and early 1960s (McGinnis, Allison, and Long 1982). Three variables are measured:

NIH	Total NIH obligations to the university in 1964 in millions of dollars.
DOCS	Number of biochemistry doctorates awarded during the period.

PDOC Number of doctorates who got postdoctoral training.

The aim is to estimate a logit model predicting the probability that a doctorate will receive postdoctoral training.

Here's the SAS code to read in the data:

```
DATA nihdoc;
  INPUT nih docs pdoc;
  DATALINES;
.5 8 1
.5 9 3
.835 16 1
.998 13 6
1.027 8 2
2.036 9 2
2.106 29 10
2.329 5 2
2.523 7 5
2.524 8 4
2.874 7 4
3.898 7 5
4.118 10 4
4.130 5 1
4.145 6 3
4.242 7 2
4.280 9 4
4.524 6 1
4.858 5 2
4.893 7 2
4.944 5 4
5.279 5 1
5.548 6 3
5.974 5 4
6.733 6 5
7 12 5
9.115 6 2
9.684 5 3
12.154 8 5
13.059 5 3
13.111 10 8
13.197 7 4
13.433 86 33
13.749 12 7
14.367 29 21
14.698 19 5
```

```
15.440 10 6
17.417 10 8
18.635 14 9
21.524 18 16
;
```

We then specify a logit model by using LOGISTIC with the *events/trials* syntax:

```
PROC LOGISTIC DATA=nihdoc;
  MODEL pdoc/docs=nih / SCALE=NONE;
RUN;
```

In Output 4.14, we see that there is a highly significant effect of NIH obligations. Each one-million dollar increase is associated with a 7.6% increase in the odds that a graduate will pursue postdoctoral training. But the model doesn't fit well. The deviance is nearly 70% larger than the degrees of freedom, with a p-value less than .01.

Output 4.14 PROC LOGISTIC Results for Doctorate Data

Deviance and Pearson Goodness-of-Fit Statistics				
Criterion	**Value**	**DF**	**Value/DF**	**Pr > ChiSq**
Deviance	64.1642	38	1.6885	0.0050
Pearson	61.0206	38	1.6058	0.0103

Model Fit Statistics			
Criterion	**Intercept Only**	**Intercept and Covariates**	**With Constant**
AIC	636.720	616.595	167.501
SC	640.849	624.853	175.759
-2 Log L	634.720	612.595	163.501

Testing Global Null Hypothesis: BETA=0			
Test	**Chi-Square**	**DF**	**Pr > ChiSq**
Likelihood Ratio	22.1252	1	<.0001
Score	21.8407	1	<.0001
Wald	21.2380	1	<.0001

Analysis of Maximum Likelihood Estimates					
Parameter	DF	Estimate	Standard Error	Wald Chi-Square	Pr > ChiSq
Intercept	1	-0.7870	0.1748	20.2718	<.0001
nih	1	0.0729	0.0158	21.2380	<.0001

Odds Ratio Estimates			
Effect	Point Estimate	95% Wald Confidence Limits	
nih	1.076	1.043	1.110

What can we do about this? Well, we can get a saturated model by deleting NIH and putting in dummy variables for 39 of the 40 universities. That wouldn't be very informative, however—we might just as well look at the percentages receiving postdoctoral training across the 40 universities. Because there's only one independent variable, we don't have the option of including interactions, but we can allow for nonlinearities by including powers of NIH in the model. I tried a squared term but that didn't help at all. The addition of a cubed term got the *p*-value up to .032, but further powers didn't bring any improvement. (I tried up to the 7th power.) In short, any polynomial model with enough terms to fit would be so complicated that it would have little appeal over the saturated model.

However, it's quite possible that the lack of fit is due not to departures from linearity in the effect NIH funding but to a lack of independence in the observations. There are many characteristics of biochemistry departments besides NIH funding that may have some impact on whether their graduates seek and get postdoctoral training. Possibilities include the prestige of the department, whether the department is in an agricultural school or a medical school, and the age of the department. Omitting these variables from the model could induce a residual correlation among the observations: People from the same department tend to have the same outcome because they share a common environment. This lack of independence will produce what is called *extra-binomial variation*—the variance of the dependent variable will be greater than what is expected under the assumption of a binomial distribution. Besides producing a large deviance, extra-binomial variation can result in underestimates of the standard errors and overestimates of the chi-square statistics.

One approach to this problem is to adjust the chi-squares and test statistics, leaving the coefficient estimates unchanged. The adjustment is quite simple: Take the ratio of the goodness-of-fit chi-square to its degrees of freedom, and then divide all the individual chi-squares by that ratio. Equivalently, take the square root of the ratio and multiply all the standard errors by that number. This adjustment, which can be based either on the Pearson chi-square or the deviance, is easily implemented in LOGISTIC by changing the SCALE option. SCALE=D makes the adjustment with the deviance chi-square and SCALE=P uses the Pearson chi-square.

Output 4.15 shows results produced by the SCALE=D option. The coefficient (and odds ratio) are identical to what we saw in Output 4.14. However, we are informed that "the covariance matrix has been multiplied by the heterogeneity factor…1.68853", which is the deviance divided by the degrees of freedom. As a consequence, all the chi-squares and all the model fit statistics in Output 4.15 are divided by this number. Thus, the Wald chi-square for NIH declines from 21.2 to 12.56, which is considerably smaller but still highly significant. The standard errors are multiplied by the square root of 1.68853. Because the deviance and the Pearson chi-square are pretty close for this data, switching to the SCALE=P option wouldn't make much difference.

Output 4.15 Doctorate Model with Overdispersion Adjustment

Deviance and Pearson Goodness-of-Fit Statistics				
Criterion	**Value**	**DF**	**Value/DF**	**Pr > ChiSq**
Deviance	64.1642	38	1.6885	0.0050
Pearson	61.0206	38	1.6058	0.0103

Model Fit Statistics			
Criterion	**Intercept Only**	**Intercept and Covariates**	**With Constant**
AIC	377.901	366.798	100.830
SC	382.030	375.056	109.088
-2 Log L	375.901	362.798	96.830

Testing Global Null Hypothesis: BETA=0			
Test	**Chi-Square**	**DF**	**Pr > ChiSq**
Likelihood Ratio	13.1032	1	0.0003
Score	12.9347	1	0.0003
Wald	12.5778	1	0.0004

Analysis of Maximum Likelihood Estimates					
Parameter	DF	Estimate	Standard Error	Wald Chi-Square	Pr > ChiSq
Intercept	1	-0.7870	0.2271	12.0056	0.0005
nih	1	0.0729	0.0206	12.5778	0.0004

Odds Ratio Estimates			
Effect	Point Estimate	95% Wald Confidence Limits	
nih	1.076	1.033	1.120

PROC LOGISTIC offers an alternative overdispersion correction proposed by Williams (1982) that modifies the coefficients as well as the standard errors and chi-squares. Based on the method of quasi-likelihood, these coefficients should be more statistically efficient than the conventional estimates; that is, they should have smaller true standard errors. The SAS code for implementing this correction is

```
PROC LOGISTIC DATA=nihdoc;
  MODEL pdoc/docs=nih / SCALE=WILLIAMS;
RUN;
```

Results are shown in Output 4.16. Williams' method uses iterative reweighting of the observations, and this is reflected in much of the reported output. In the preliminary output, we see the weight function used at the final iteration, along with the sum of the weights. This sum can be thought of as the effective sample size after correction for overdispersion. Although both Pearson and deviance chi-squares are reported, there is a warning that they are not to be used to assess the fit of the model. Both the coefficient and the chi-square for NIH are a little larger than they were under the simpler SCALE=D correction in Output 4.15. Note that when the group size variable (DOCS in this example) has the same value for all groups, Williams' method does not alter the conventional coefficient estimates, and the standard error correction is the same as the simpler adjustment using Pearson's chi-square.

Output 4.16 Doctorate Model with Williams' Adjustment for Overdispersion

Model Information	
Data Set	WORK.NIHDOC
Response Variable (Events)	pdoc
Response Variable (Trials)	docs
Weight Variable	1 / (1 + 0.042183 * (docs - 1))

Model Information	
Model	binary logit
Optimization Technique	Fisher's scoring

Number of Observations Read	40
Number of Observations Used	40
Sum of Frequencies Read	459
Sum of Frequencies Used	459
Sum of Weights Read	276.8967
Sum of Weights Used	276.8967

Response Profile			
Ordered Value	**Binary Outcome**	**Total Frequency**	**Total Weight**
1	Event	216	133.24419
2	Nonevent	243	143.65250

Deviance and Pearson Goodness-of-Fit Statistics				
Criterion	**Value**	**DF**	**Value/DF**	**Pr > ChiSq**
Deviance	39.8872	38	1.0497	0.3862
Pearson	37.9995	38	1.0000	0.4695

Number of events/trials observations: 40

Note: Since the Williams method was used to accommodate overdispersion, the Pearson chi-squared statistic and the deviance can no longer be used to assess the goodness of fit of the model.

Model Fit Statistics			
Criterion	**Intercept Only**	**Intercept and Covariates**	**With Constant**
AIC	385.469	370.396	114.478
SC	389.598	378.654	122.736
-2 Log L	383.469	366.396	110.478

Testing Global Null Hypothesis: BETA=0			
Test	**Chi-Square**	**DF**	**Pr > ChiSq**
Likelihood Ratio	17.0734	1	<.0001
Score	16.7876	1	<.0001
Wald	16.1391	1	<.0001

Analysis of Maximum Likelihood Estimates					
Parameter	DF	Estimate	Standard Error	Wald Chi-Square	Pr > ChiSq
Intercept	1	-0.7521	0.2084	13.0309	0.0003
nih	1	0.0829	0.0206	16.1391	<.0001

Odds Ratio Estimates			
Effect	Point Estimate	95% Wald Confidence Limits	
nih	1.086	1.043	1.131

What makes the doctorate example different from the earlier examples in this chapter is that the individuals are grouped into naturally occurring clusters, in this case, university departments. It's reasonable to suppose that individuals in the same department are *not* independent—not only can they influence each other, but they are also exposed to many common factors that may produce the same outcome. In most contingency table analyses, on the other hand, individuals are grouped together merely because they have the same values on some discrete variable. In such cases, there is usually no reason to think that the observations within groups are anything other than independent.

When independence is presumed, the correct strategy for dealing with overdispersion is to elaborate the model until you find a version that does fit the data, not to casually invoke the overdispersion options. If, after diligent investigation, you cannot come up with a reasonably parsimonious model with an acceptable fit to the data, then there may be some value in correcting the standard errors and test statistics for any remaining overdispersion. But be aware that the correction for overdispersion always produces chi-square statistics that are lower than they would be without the correction.

Note also that overdispersion does *not* arise from heterogeneity *within* groups. The problem stems from differences *between* groups, differences that are not fully described by the measured variables. Another way to deal with overdispersion is to postulate a random effect for departments that accounts for the overdispersion. This can be done with PROC GLIMMIX or PROC NLMIXED. In Chapter 8, we'll see how to use these procedures to analyze individual-level data when there is clustering and a lack of independence within clusters. See Allison (1987) for still other ways to approach the overdispersion problem for these data.

It's also worth noting that the "badness of fit" for the NIH data comes primarily from a small number of observations. The graph in Output 4.17 shows the change in the deviance with inclusion vs. deletion of each observation. This plot is one of several that is produced when the option PLOTS(LABEL UNPACK)=LEVERAGE is used on the PROC statement. We see that observations 33 and 36 have deviance differences that are more than 7.5. Observation 33 is the largest department with 86 doctorates and 33 postdocs. Observation 36 had 19 doctorates and 5 postdocs. When the model is refit after deleting both of these observations, the deviance falls to 37.9 with 36 degrees of freedom, for a *p*-value of .38. While this result is not inconsistent with the hypothesis of overdispersion, it does suggest that special attention be paid to these two departments to determine how they differ from others.

Output 4.17 Deviance Differences by Observation Number

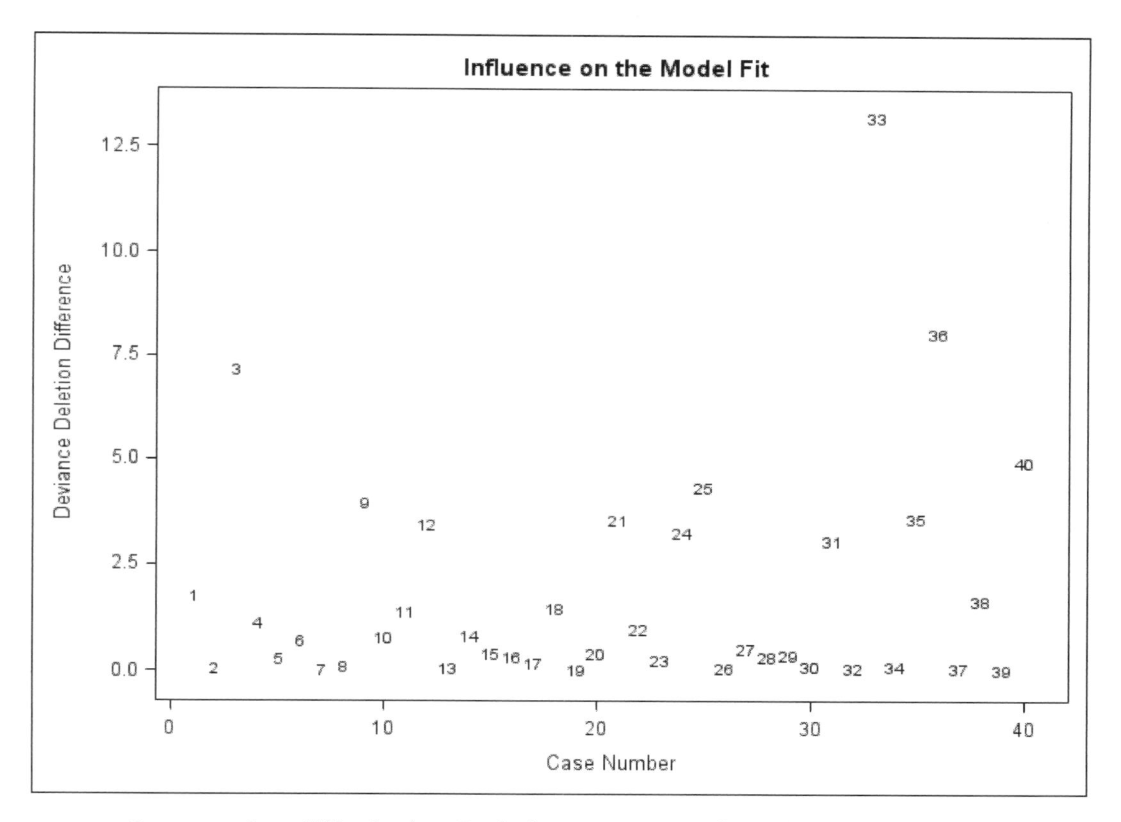

One way they differ is size. Both departments graduated many more doctorates than the median number of 8. Perhaps we can improve the fit by including DOCS itself as an independent variable in the model:

```
PROC LOGISTIC DATA=nihdoc;
  MODEL pdoc/docs=nih docs/ SCALE=NONE;
RUN;
```

As shown in Output 4.18, this modification does produce a much better fitting model, with a deviance *p*-value of .07. Departmental size has a highly significant, negative effect on the probability that a new doctorate will get a postdoctoral fellowship. In general, the most desirable way to deal with overdispersion is to incorporate covariates that account for differences among the groups. But that may not always be possible.

Output 4.18 Doctorate Model with DOCS as a Covariate

Deviance and Pearson Goodness-of-Fit Statistics				
Criterion	Value	DF	Value/DF	Pr > ChiSq
Deviance	50.3161	37	1.3599	0.0709
Pearson	47.8917	37	1.2944	0.1083

Analysis of Maximum Likelihood Estimates					
Parameter	DF	Estimate	Standard Error	Wald Chi-Square	Pr > ChiSq
Intercept	1	-0.6662	0.1786	13.9103	0.0002
nih	1	0.0988	0.0178	30.6586	<.0001
docs	1	-0.0131	0.00358	13.3971	0.0003

Chapter 5

Multinomial Logit Analysis

5.1 Introduction

Binary logistic regression is ideal when your dependent variable has two categories, but what if it has three or more? In some cases, it may be reasonable to collapse categories so that you have only two, but that strategy inevitably involves some loss of information. In other cases, collapsing categories could seriously obscure what you're trying to study. Suppose you want to estimate a model predicting whether newly registered voters choose to register as Democrats, Republicans, or Independents. Combining any two of these outcomes could lead to seriously misleading conclusions.

How you deal with such situations depends somewhat on the nature of the outcome variable and the goal of the analysis. If the categories of the dependent variable have an inherent ordering, the methods in the next chapter should do the job. If there's no inherent ordering and the goal is a model in which *characteristics of the outcome categories* predict choice of category, then the discrete-choice model of Chapter 7 is probably what you need. In this chapter we consider models for unordered categories where the predictor variables are characteristics of the individual, and possibly the individual's environment. For example, we could estimate a model predicting party choice of newly registered voters based on information about the voter's age, income, and years of schooling. We might also include information about the precinct in which the voter is registering.

The model is called the multinomial logit model because the probability distribution for the outcome variable is assumed to be a multinomial rather than a binomial distribution. In SAS documentation, it's often referred to as the generalized logit model. Because the multinomial logit model can be rather complicated to interpret when the outcome variable has many categories, I'll begin with the relatively simple case of a three-category outcome.

5.2 Example

As in the binomial case, let's start with a real example. Several years ago, I did a survey of 195 undergraduates at the University of Pennsylvania in order to study the effects of parenting styles on altruistic behavior. One of the questions was, "If you found a wallet on the street, would you (1) keep the wallet and the money, (2) keep the money and return the wallet, or (3) return both the wallet and the money." The distribution of responses for the WALLET variable was

Value	Frequency	
1	24	keep both
2	50	keep the money
3	121	return both

Possible explanatory variables are

MALE	1= male, 0 = female
BUSINESS	1=enrolled in business school, 0=otherwise
PUNISH	A variable describing whether the student was physically punished by parents at various ages:

1 = punished in elementary school but not middle or high school

2 = punished in elementary and middle school but not high school

3 = punished at all three levels

EXPLAIN "When you were punished, did your parents generally explain why what you did was wrong?"

1 = almost always

0 = sometimes or never

In the next section we construct a multinomial logit model for the dependence of WALLET on the explanatory variables.

5.3 A Model for Three Categories

First, some notation. Define

p_{i1} = the probability that WALLET=1 for person i

p_{i2} = the probability that WALLET=2 for person i

p_{i3} = the probability that WALLET=3 for person i

Let $\mathbf{x}_i$ be a column vector of explanatory variables for person i:

$$\mathbf{x}_i = [1 \;\; x_{i1} \;\; x_{i2} \;\; x_{i3} \;\; x_{i4}]'$$

where the x's refer to MALE, BUSINESS, PUNISH, and EXPLAIN. If you're not comfortable with vectors, you can just think of $\mathbf{x}_i$ as a single explanatory variable. In order to generalize the logit model to this three-category case, it's tempting to specify three binary logit models, one for each outcome:

$$\log\left(\frac{p_{i1}}{1-p_{i1}}\right) = \boldsymbol{\beta}_1 \mathbf{x}_i$$

$$\log\left(\frac{p_{i2}}{1-p_{i2}}\right) = \boldsymbol{\beta}_2 \mathbf{x}_i$$

$$\log\left(\frac{p_{i3}}{1-p_{i3}}\right) = \boldsymbol{\beta}_3 \mathbf{x}_i$$

where the $\boldsymbol{\beta}$'s are row vectors of coefficients. This turns out to be an unworkable approach, however. Because $p_{i1}+p_{i2}+p_{i3}=1$, these three equations are inconsistent. If the first two equations are correct, for example, the third cannot be correct. Instead, we formulate the model as follows:

$$\log\left(\frac{p_{i1}}{p_{i3}}\right) = \boldsymbol{\beta}_1\mathbf{x}_i$$

$$\log\left(\frac{p_{i2}}{p_{i3}}\right) = \boldsymbol{\beta}_2\mathbf{x}_i$$

$$\log\left(\frac{p_{i1}}{p_{i2}}\right) = \boldsymbol{\beta}_3\mathbf{x}_i$$

These equations are mutually consistent, and one is redundant. For example, the third equation can be obtained from the first two. Using properties of logarithms, we have

$$\begin{aligned}\log\left(\frac{p_{i1}}{p_{i2}}\right) &= \log\left(\frac{p_{i1}}{p_{i3}}\right) - \log\left(\frac{p_{i2}}{p_{i3}}\right) \\ &= \boldsymbol{\beta}_1\mathbf{x}_i - \boldsymbol{\beta}_2\mathbf{x}_i \\ &= (\boldsymbol{\beta}_1 - \boldsymbol{\beta}_2)\mathbf{x}_i\end{aligned}$$

which implies that $\boldsymbol{\beta}_3 = \boldsymbol{\beta}_1 - \boldsymbol{\beta}_2$. Solving for the three probabilities, we get

$$p_{i1} = \frac{e^{\boldsymbol{\beta}_1\mathbf{x}_i}}{1 + e^{\boldsymbol{\beta}_1\mathbf{x}_i} + e^{\boldsymbol{\beta}_2\mathbf{x}_i}}$$

$$p_{i2} = \frac{e^{\boldsymbol{\beta}_2\mathbf{x}_i}}{1 + e^{\boldsymbol{\beta}_1\mathbf{x}_i} + e^{\boldsymbol{\beta}_2\mathbf{x}_i}}$$

$$p_{i3} = \frac{1}{1 + e^{\boldsymbol{\beta}_1\mathbf{x}_i} + e^{\boldsymbol{\beta}_2\mathbf{x}_i}}\ .$$

Because the three numerators sum to the common denominator, we immediately verify that the three probabilities sum to 1.

As with the binary logit model, the most widely used method of estimation is maximum likelihood. I won't go through the derivation, but it's very similar to the binary case. Again, the Newton-Raphson algorithm is generally used to get the maximum likelihood estimates. This can be accomplished in any of the following procedures: LOGISTIC, SURVEYLOGISTIC, CATMOD, or GLIMMIX. This chapter focuses exclusively on LOGISTIC because it's already familiar, and it's somewhat simpler to use and interpret.

5.4 Estimation with PROC LOGISTIC

Here's the SAS code to estimate a multinomial logit model for the wallet data:

```
PROC LOGISTIC DATA=wallet;
  MODEL wallet = male business punish explain / LINK=GLOGIT;
RUN;
```

The unordered multinomial model is invoked by the LINK=GLOGIT option. If this option is omitted, LOGISTIC estimates the *cumulative* logit, which assumes that the response levels are ordered. The cumulative model is described in the next chapter. For now, I'm treating PUNISH as a quantitative variable, but later I'll explore a categorical coding. Results are shown in Output 5.1.

Output 5.1 PROC LOGISTIC Output for Wallet Data

Model Information	
Data Set	MY.WALLET
Response Variable	wallet
Number of Response Levels	3
Model	generalized logit
Optimization Technique	Newton-Raphson

Number of Observations Read	195
Number of Observations Used	195

Logits modeled use wallet=3 as the reference category.

Model Convergence Status
Convergence criterion (GCONV=1E-8) satisfied.

Model Fit Statistics		
Criterion	Intercept Only	Intercept and Covariates
AIC	356.140	321.586
SC	362.686	354.316
-2 Log L	352.140	301.586

Testing Global Null Hypothesis: BETA=0			
Test	Chi-Square	DF	Pr > ChiSq
Likelihood Ratio	50.5540	8	<.0001
Score	51.6357	8	<.0001
Wald	38.8409	8	<.0001

Type 3 Analysis of Effects			
Effect	DF	Wald Chi-Square	Pr > ChiSq
male	2	12.2818	0.0022
business	2	4.6858	0.0961
punish	2	10.9196	0.0043
explain	2	9.7651	0.0076

Analysis of Maximum Likelihood Estimates						
Parameter	wallet	DF	Estimate	Standard Error	Wald Chi-Square	Pr > ChiSq
Intercept	1	1	-3.4712	0.8440	16.9139	<.0001
Intercept	2	1	-1.2917	0.6073	4.5237	0.0334
male	1	1	1.2672	0.5546	5.2207	0.0223
male	2	1	1.1699	0.3715	9.9178	0.0016
business	1	1	1.1804	0.5487	4.6281	0.0315
business	2	1	0.4179	0.4233	0.9747	0.3235
punish	1	1	1.0817	0.3336	10.5162	0.0012
punish	2	1	0.1957	0.2889	0.4591	0.4981
explain	1	1	-1.6006	0.5455	8.6083	0.0033
explain	2	1	-0.8040	0.4034	3.9722	0.0463

Odds Ratio Estimates				
Effect	wallet	Point Estimate	95% Wald Confidence Limits	
male	1	3.551	1.197	10.530
male	2	3.222	1.556	6.673
business	1	3.256	1.111	9.542
business	2	1.519	0.662	3.482
punish	1	2.950	1.534	5.671
punish	2	1.216	0.690	2.143
explain	1	0.202	0.069	0.588
explain	2	0.448	0.203	0.987

Let's look first at the "Type 3 Analysis of Effects" table. Each chi-square is a test of the null hypothesis that the explanatory variable has no effect on the outcome variable. In this example, there are two degrees of freedom for each chi-square because each variable has two coefficients. So the null hypothesis is that both coefficients are 0. BUSINESS has a *p*-value of .096, but the other three variables are all significant at beyond the .01 level.

In the next table of Output 5.1, “Analysis of Maximum Likelihood Estimates,” we find the coefficient estimates and associated statistics. If the dependent variable has *K* categories, LOGISTIC estimates *K*–1 equations. So with a 3-category outcome there are two equations. For each variable, the first coefficient (labeled 1 in the WALLET column) is for the first equation, and the second coefficient (labeled 2 in the WALLET column) is for the second equation. Each equation is a contrast between a given category and a reference category. The default reference category is the highest value of the dependent variable. In this case, it’s WALLET=3, return the wallet and money. So the first equation is a model for category 1 vs. category 3, which is keeping both vs. returning both. The second equation is a model for 2 vs. 3, which is keeping the money and returning the wallet vs. returning both.

To make this more clear, I often find it helpful to reorganize the LOGISTIC output into separate equations, as shown in Table 5.1. The first column corresponds to equation 1, and the second column corresponds to equation 2. The third column, which is an equation predicting category 1 vs. category 2, can be obtained by simply subtracting the numbers in column 2 from the numbers in column 1. Alternatively, you can get the numbers in the third column by re-running the program after changing the reference category of WALLET to “keep the money” rather than “return both.” This is accomplished with the following MODEL statement:

```
MODEL wallet(REF='2') = male business punish explain /
LINK=GLOGIT;
```

With the new reference category, all the coefficients labeled 1 (in the WALLET column) correspond to those in the third column of Table 5.1. By doing it this way, you also get *p*-values for all the coefficients.

Table 5.1 Reorganized Output from PROC LOGISTIC

	Keep Both vs. Return Both	Keep One vs. Return Both	Keep Both vs. Keep One
INTERCEPT	–3.47**	–1.29*	–2.18*
MALE	1.27*	1.17*	.10
BUSINESS	1.18*	.42	.76
PUNISH	1.08**	.20	.89*
EXPLAIN	–1.60**	–.80*	–.80

*p<.05, **p<.01

Examining the first column, we see that all the variables have significant effects on keeping both vs. returning both. The corresponding odds ratios (in Output 5.1) may be interpreted just like odds ratios in a binary logit model, except that they describe *conditional* odds. For example, the odds ratio for MALE is 3.56. We can then say that the odds that a male will keep both rather than return both (conditional on not keeping just the money) are about 3.5 times the odds for females. Similarly, the odds ratio for BUSINESS is 3.25. So we may say that students in the business school have an odds of keeping both vs. returning both that is about 3.25 times the odds for those in other schools, again conditional on not keeping just the money. For PUNISH we have an odds ratio of 2.95, implying that each 1-level increase in that variable multiplies the odds of keeping both vs. returning both by about 3. Finally, students whose parents explained their punishments had odds of keeping both vs. returning both that were only one-fifth the odds for those parents who did not explain.

The coefficients in the second column have the same sign as those in the first but are generally much smaller. That's not surprising because the behaviors being compared are less extreme. We still see a big effect of gender. The odds that a male will keep the money vs. returning both are more than three times the odds for females (conditional on not keeping both). While the coefficient for EXPLAIN has been cut in half, the odds for those whose parents did *not* explain are still more than double the odds for those whose parents did explain.

The third column has a dual interpretation. The coefficients represent the effect of the explanatory variables on being in category 1 vs. category 2, conditional on not being in category 3. But they also represent the difference between the coefficients in the first two columns. PUNISH is the only variable whose coefficient declined significantly from column 1 to column 2. Equivalently, PUNISH is the only variable having a significant effect on keeping both vs. keeping the money only. Each one-level increase of PUNISH more than doubles the odds of keeping both vs. keeping the money only.

With a TEST statement, we can test the null hypothesis that *all* the coefficients in the first column are identical to the corresponding coefficients in the second column:

```
PROC LOGISTIC DATA=wallet;
  MODEL wallet = male business punish explain / LINK=GLOGIT;
  TEST male_1=male_2, business_1=business_2,
    punish_1=punish_2, explain_1=explain_2;
RUN;
```

As this program shows, we can refer to the coefficient for a variable in the first equation by putting _1 after the variable name in the TEST statement. To refer to a coefficient in the second equation, we append _2. The value after the underscore must be the actual value that the variable takes on, which could be a character value. Thus, if WALLET had values of A, B, and C instead of 1, 2, and 3, the TEST statement would have MALE_A=MALE_B, and so on.

The TEST statement tests the null hypothesis that all four pairs of coefficients are equal. It produces a Wald chi-square of 12.2844 with 4 degrees of freedom, yielding a *p*-value of .0154. We conclude that the coefficients estimated in columns 1 and 2 of Table 5.1 are *not* the same. If this test had produced a high *p*-value, it would suggest that categories 1 and 2 of WALLET could be combined into a single category.

How do we know if the model fits the data? The best way to approach this question is to consider the ways in which the model might be wrong. First, there could be interactions among the predictors in their effects on WALLET. And, second, treating PUNISH as a quantitative variable could be incorrect. That is, the effect of going from category 1 of PUNISH to category 2 could be different from the effect of going from 2 to 3.

Since all the predictors are discrete with a small number of values, we can test for both of these possibilities by requesting goodness-of-fit statistics, using the AGGREGATE and SCALE=NONE options:

```
PROC LOGISTIC DATA=wallet;
  MODEL wallet = male business punish explain / LINK=GLOGIT
    AGGREGATE SCALE=NONE;
  OUTPUT OUT=predicted PREDPROBS=I;
RUN;
```

I'll discuss the OUTPUT statement in a moment. The deviance and Pearson chi-square statistics, shown in Output 5.2, have very high *p*-values, suggesting that the model fits the data well.

Output 5.2 Goodness-of-Fit Statistics for Multinomial Logit Model

Deviance and Pearson Goodness-of-Fit Statistics				
Criterion	**Value**	**DF**	**Value/DF**	**Pr > ChiSq**
Deviance	31.9450	36	0.8874	0.6619
Pearson	26.4726	36	0.7353	0.8770

Number of unique profiles: 23

Notice that there are 23 "unique profiles," which is the number of unique combinations of the values of the predictor variables. With 195 total cases, we have an average of 8.5 cases per profile, which is reassuring. These tests tend to produce inaccurate *p*-values when the number of cases per profile gets too small. But to really evaluate the accuracy of these *p*-values, we need to examine the expected frequencies in each cell of the multiway contingency table.

That's where the OUTPUT statement comes in. It produces a data set (named PREDICTED) containing the predicted probabilities of being in each category of the dependent variable. The data set has one record for each of the 195 cases. The PREDPROBS=I option requests that three new variables be generated: IP_1 is the predicted probability of being in category 1, IP_2 is the predicted probability of being in category 2, and IP_3 is the predicted probability of being in category 3. If we sum these predicted probabilities within each profile, we get the expected frequencies. We can accomplish that with PROC TABULATE. Here's the code:

```
PROC TABULATE DATA=predicted;
  CLASS male business punish explain;
  var IP_1 IP_2 IP_3;
  TABLE male*business*punish*explain, IP_1 IP_2 IP_3;
RUN;
```

Output 5.3 Expected Frequencies for Multinomial Logit Model

male	business	punish	explain	Individual Probability: wallet=1 Sum	Individual Probability: wallet=2 Sum	Individual Probability: wallet=3 Sum
0	0	1	0	0.77	2.81	8.42
			1	0.79	6.40	42.81
		2	0	1.13	1.70	4.17
			1	0.31	1.03	5.66
		3	0	1.74	1.08	2.18
			1	0.35	0.48	2.17
	1	1	0	0.17	0.28	0.55
			1	0.23	0.88	3.88
		2	0	0.35	0.25	0.40
			1	0.37	0.57	2.06
		3	0	1.79	0.52	0.69
1	0	1	0	1.22	4.03	3.75
			1	1.70	12.46	25.85
		2	0	2.35	3.20	2.45
			1	0.33	0.99	1.69
		3	0	1.57	0.88	0.55
			1	0.75	0.94	1.31
	1	1	0	1.43	2.21	1.35
			1	1.76	6.02	8.22
		2	0	1.02	0.65	0.33
			1	1.00	1.41	1.59
		3	0	1.46	0.38	0.16
			1	1.42	0.82	0.76

As seen in Output 5.3, many of the expected frequencies are less than 1, and most of them are less than 2. This does not augur well for the performance of the goodness of fit statistics.

A better approach is to directly test for the presence of interactions by including them in the model. As we have seen, LOGISTIC allows multiplicative terms to be included directly on the MODEL statement. Using the short-hand notation described in the previous chapter, the following model contains all six 2-way interactions as well as the "main effects":

```
MODEL wallet = male | business | punish | explain @2
   / LINK=GLOGIT;
```

Results in Output 5.4 give little evidence for interactions—the smallest *p*-value is .22. To test the hypothesis that *all* the 2-way interactions are 0, we can take the difference in the likelihood ratio chi-squares (for testing the global null hypothesis: BETA=0) for this model and the one in Output 5.2. That yields a chi-square of 9.17 with 12 degrees of freedom, which is far from statistically significant. It's possible that there could be higher-order interactions. But with no 2-way interactions, that seems unlikely.

Output 5.4 Type 3 Table for Two-Way Interaction Model

Type 3 Analysis of Effects			
Effect	DF	Wald Chi-Square	Pr > ChiSq
male	2	0.7625	0.6830
business	2	0.9113	0.6340
male*business	2	2.4937	0.2874
punish	2	3.3584	0.1865
male*punish	2	0.3392	0.8440
business*punish	2	0.9822	0.6119
explain	2	0.1815	0.9133
male*explain	2	1.2697	0.5300
business*explain	2	0.8757	0.6454
punish*explain	2	2.9969	0.2235

The other issue is whether it's appropriate to treat PUNISH as a quantitative variable with a linear effect on the log of the odds. There are two ways to address that question. One way is to include the square of PUNISH in the model (by putting PUNISH*PUNISH on the MODEL statement). Since PUNISH only has three values, that exhausts all the degrees of freedom for this variable. The squared term has a *p*-value of .98 (table not shown), giving no evidence for departure from linearity. The other method is to refit the model with PUNISH as CLASS variable. Then take the difference between the likelihood ratio chi-squares for this model and for the model with PUNISH as a quantitative variable. That produces a chi-square of only .03 with two degrees of freedom, which is far from statistically significant. So all our tests lead to the conclusion that our original model (in Output 5.1) is consistent with the data.

The EFFECTPLOT statement can be very helpful in visualizing the effect of a variable in a multinomial logit model. The following program shows how to get a plot of the predicted probabilities of falling into each of the three outcome categories as a function of PUNISH:

```
PROC LOGISTIC DATA=wallet;
  MODEL wallet = male business punish explain / LINK=GLOGIT;
  EFFECTPLOT FIT(X=punish) / NOOBS NOLIMITS GRIDSIZE=3;
RUN;
```

For this plot, the other three variables are held at their means. The NOOBS option suppresses the inclusion of the data point markers in the graph. The NOLIMITS option suppresses the 95% confidence bands around the plots. The GRIDSIZE=3 option is included because PUNISH has only three equally spaced values (1, 2, 3). Without this option, the graphs would consist of smooth curves rather than the more appropriate broken lines that are shown Output 5.5.

Output 5.5 Plot of Predicted Probabilities as a Function of PUNISH

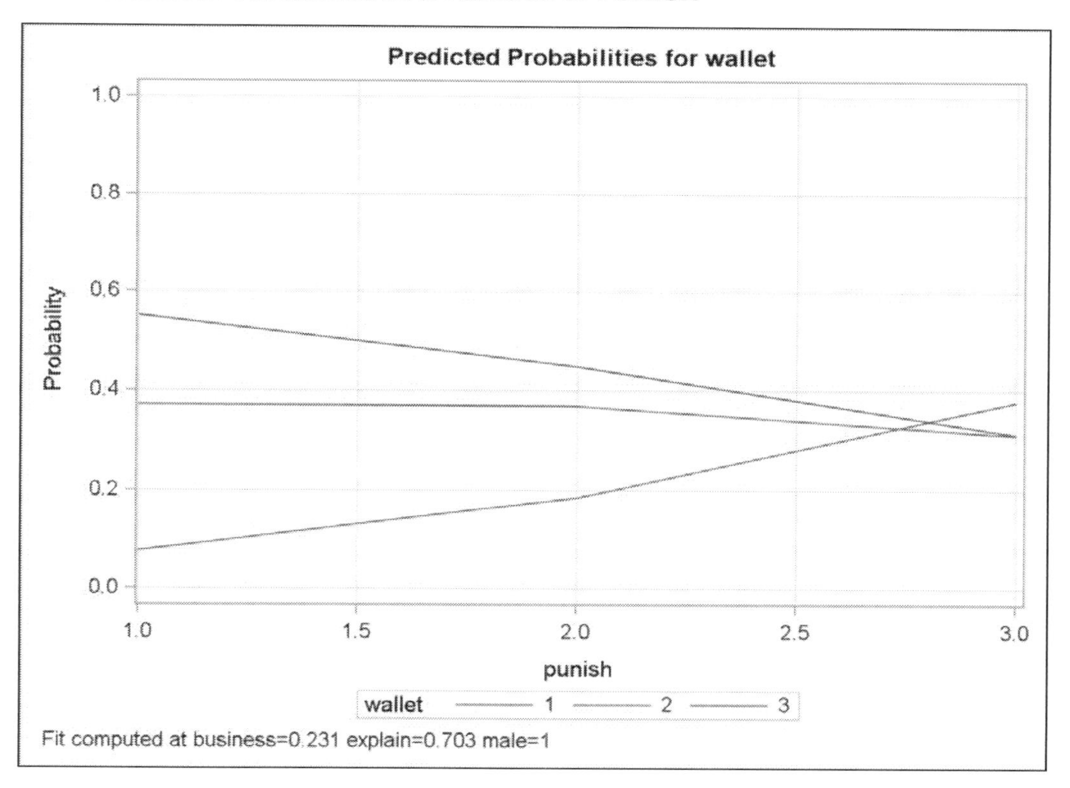

5.5 Estimation with a Binary Logit Procedure

As we've just seen, the multinomial logit model can be interpreted as a set of binary logit equations, each equation corresponding to a comparison between two of the categories of the dependent variable. It turns out that you can legitimately estimate the multinomial logit model by running a set of binary logit models (Begg and Gray 1984). For each two-category contrast, observations that fall into other categories are excluded. Here's how to do it for the three columns of Table 5.1:

```
PROC LOGISTIC DATA=wallet;
  WHERE wallet NE 2;
  MODEL wallet=male business punish explain;
PROC LOGISTIC DATA=wallet;
  WHERE wallet NE 1;
  MODEL wallet=male business punish explain;
PROC LOGISTIC DATA=wallet;
  WHERE wallet NE 3;
  MODEL wallet=male business punish explain;
RUN;
```

In the WHERE statements, NE means "not equal to." Because WALLET has only three categories, removing any one of them reduces the model to a binary logit model. By default, the reference category of the dependent variable in each logit model is the higher value. So the first model predicts a 1 rather than a 3, the second model predicts a 2 rather than a 3, and the third model predicts a 1 rather than a 2.

Results are shown in Table 5.2. Comparing this with Table 5.1, we see that the coefficients are very similar but not identical. The significance levels are also pretty much the same, although the *p*-value for BUSINESS in the KEEP BOTH VS. RETURN BOTH column is slightly below .05 in Table 5.1 and slightly above in Table 5.2.

Table 5.2 Multinomial Estimates by Using Binary Logits

	Keep Both vs. Return Both	Keep One vs. Return Both	Keep Both vs. Keep One
INTERCEPT	–3.53**	–1.33*	–2.28*
MALE	1.42*	1.14*	.18
BUSINESS	1.07	.38	.79
PUNISH	1.09**	.23	.87*
EXPLAIN	–1.63**	–.77*	–.68

*$p<.05$, **$p<.01$

The coefficients in Table 5.2 have no more bias than the ones in Table 5.1 (in technical terms, they have the same probability limit). But the binary estimates in Table 5.2 are less efficient—they have somewhat larger true standard errors. Furthermore, unlike Table 5.1, the third column of Table 5.2 is *not* exactly equal to the difference between the first two columns. Finally, estimation by binary logits won't give you a global test of whether *all* the coefficients for a particular variable are equal to 0, as do the tests in the Type 3 table of Output 5.1.

What's the point of this exercise? Well, occasionally you may find yourself in a situation where it's more convenient to do binary estimation than multinomial estimation. One possible advantage in doing it this way is that you can have different sets of variables in different equations. Another advantage is that there many options (like the Firth method described in Chapter 3) that are available for binary logit but not for multinomial logit. But I think the main payoff is conceptual. When I first learned this equivalence, it really helped me understand that the multinomial model is built up of binomial models. The simultaneous estimation procedure is just a technical refinement.

5.6 General Form of the Model

To this point, we've been dealing with just three categories for the dependent variable. Now let's generalize the model to J categories, with the running index j=1, …, J. Let p_{ij} be the probability that individual i falls into category j. The model is then

$$\log\left(\frac{p_{ij}}{p_{iJ}}\right) = \boldsymbol{\beta}_j \mathbf{x}_i \qquad j = 1,\ldots,J-1 \qquad (5.1)$$

where $\mathbf{x}_i$ is a column vector of variables describing individual i, and $\boldsymbol{\beta}_j$ is a row vector of coefficients for category j. Note that each category is compared with the highest category J. These equations can be solved to yield

$$p_{ij} = \frac{e^{\boldsymbol{\beta}_j \mathbf{x}_i}}{1+\sum_{k=1}^{J-1} e^{\boldsymbol{\beta}_k \mathbf{x}_i}} \qquad j = 1,\ldots,J-1 \qquad (5.2)$$

Because the probabilities for all J categories must sum to 1, we have

$$p_{iJ} = \frac{1}{1+\sum_{k=1}^{J-1} e^{\boldsymbol{\beta}_k \mathbf{x}_i}}$$

After the coefficients are estimated, the logit equation for comparing any two categories j and k of the dependent variable can be obtained from

$$\log\left(\frac{p_{ij}}{p_{ik}}\right)=(\boldsymbol{\beta}_j-\boldsymbol{\beta}_k)\mathbf{x}_i \ .$$

5.7 Contingency Table Analysis

As with binary logit analysis, the multinomial logit model can be easily applied to the analysis of contingency tables. Consider Table 5.3, which was tabulated from the 1991 General Social Survey (Agresti 2002).

Table 5.3 Belief in an Afterlife, by Gender and Race

		Belief in Afterlife		
		Yes	Undecided	No
White	Female	371	49	74
	Male	250	45	71
Black	Female	64	9	15
	Male	25	5	13

The dependent variable—belief in an afterlife—has three categories, which we'll treat as unordered. Here's how to fit the multinomial logit model with LOGISTIC:

```
DATA afterlif;
  INPUT white female belief count;
  DATALINES;
1 1 1 371
1 1 2  49
1 1 3  74
1 0 1 250
1 0 2  45
1 0 3  71
0 1 1  64
0 1 2   9
0 1 3  15
0 0 1  25
```

```
0 0 2   5
0 0 3  13
;
PROC LOGISTIC DATA=afterlif;
  FREQ count;
  MODEL belief=white female / LINK=GLOGIT AGGREGATE
   SCALE=NONE;
RUN;
```

Each of the two predictor variables is coded 1 or 0. The dependent variable BELIEF is coded: 1=Yes, 2=Undecided, and 3=No. Because "No" has the highest value, it becomes the reference category for the model.

Results are shown in Output 5.6. The deviance and Pearson chi-square tests indicate no significant lack of fit. Because the only difference between this model and a "saturated" model is the interaction between race and gender, the high *p*-values for the goodness of fit statistics simply mean that there is no evidence for that interaction. All of the three chi-square tests for the "global null hypothesis" give *p*-values of around .065. So there is only marginal evidence that either of the two predictors has an effect on the response. From the Type 3 table, however, we see evidence for a gender effect but no indication of a race effect.

Output 5.6 PROC LOGISTIC Results for Afterlife Data

Response Profile

Ordered Value	belief	Total Frequency
1	1	710
2	2	108
3	3	173

Logits modeled use belief=3 as the reference category.

Deviance and Pearson Goodness-of-Fit Statistics

Criterion	Value	DF	Value/DF	Pr > ChiSq
Deviance	0.8539	2	0.4269	0.6525
Pearson	0.8609	2	0.4304	0.6502

Testing Global Null Hypothesis: BETA=0			
Test	**Chi-Square**	**DF**	**Pr > ChiSq**
Likelihood Ratio	8.7437	4	0.0678
Score	8.8498	4	0.0650
Wald	8.7818	4	0.0668

Type 3 Analysis of Effects			
Effect	**DF**	**Wald Chi-Square**	**Pr > ChiSq**
male	2	12.2818	0.0022
business	2	4.6858	0.0961
punish	2	10.9196	0.0043
explain	2	9.7651	0.0076

Analysis of Maximum Likelihood Estimates						
Parameter	**belief**	**DF**	**Estimate**	**Standard Error**	**Wald Chi-Square**	**Pr > ChiSq**
Intercept	**1**	1	0.8828	0.2426	13.2390	0.0003
Intercept	**2**	1	-0.7582	0.3614	4.4031	0.0359
white	**1**	1	0.3420	0.2370	2.0814	0.1491
white	**2**	1	0.2712	0.3541	0.5863	0.4438
female	**1**	1	0.4186	0.1713	5.9737	0.0145
female	**2**	1	0.1051	0.2465	0.1817	0.6699

Odds Ratio Estimates				
Effect	**belief**	**Point Estimate**	**95% Wald Confidence Limits**	
white	**1**	1.408	0.885	2.240
white	**2**	1.311	0.655	2.625
female	**1**	1.520	1.086	2.126
female	**2**	1.111	0.685	1.801

In the “Analysis of Maximum Likelihood Estimates” table, all of the parameters with a value of 1 in the BELIEF column pertain to the equation for yes vs. no (1 vs. 3); the parameters with a BELIEF value of 2 pertain to the equation for undecided vs. no (2 vs. 3). The gender effect seems to be entirely concentrated in the yes/no contrast. With an odds ratio of 1.52, we can say that women are about 50% more likely than men to say that they believe in an afterlife. However, the 95 percent confidence interval for the odds ratio is quite wide, ranging from 1.086 to 2.126.

5.8 Problems of Interpretation

When the dependent variable has only three categories, the multinomial logit model is reasonably easy to interpret. But as the number of categories increases, it becomes more and more difficult to tell a simple story about the results. As we've seen, the model is most naturally interpreted in terms of effects on contrasts between pairs of categories for the dependent variable. But with a 5-category dependent variable, there are 10 different possible pairs of categories. With a 10-category dependent variable, the number of pairs shoots up to 45. There's no simple solution to this problem, but Long (1997) has proposed a graphical method of interpreting the coefficients that can be very helpful for a moderate number of categories.

There's still another interpretational problem that even experienced users of multinomial logit analysis may overlook. In binary logit analysis, if a variable x has a positive coefficient, you know that every increase in x results in an increase in the probability of the designated outcome. That's not always true in the multinomial model.

Consider the following hypothetical example. Suppose the dependent variable y has three categories (1, 2, and 3) and there is a single independent variable x. When we estimate a multinomial logit model, let's suppose that we get the following two equations:

$$\log\left(\frac{p_1}{p_3}\right) = 1.0 - 2.0x$$

$$\log\left(\frac{p_2}{p_3}\right) = 1.0 - 1.0x.$$

How can we describe the effect of x on p_2? Because the coefficient of x in the second equation is negative, it's tempting to say that increases in x produce decreases in the probability of being in category 2. But if we graph the probabilities, as in Output 5.7, we see that the effect of x on p_2 is not monotonic. (This graph was produced with simulated data using the EFFECTPLOT statement.) When x is below .5, an increase in x produces an increase in p_2. When x is above .5, increases in x produce decreases in p_2.

Output 5.7 Effect of x on the Probability of Being in Each of the Three Categories

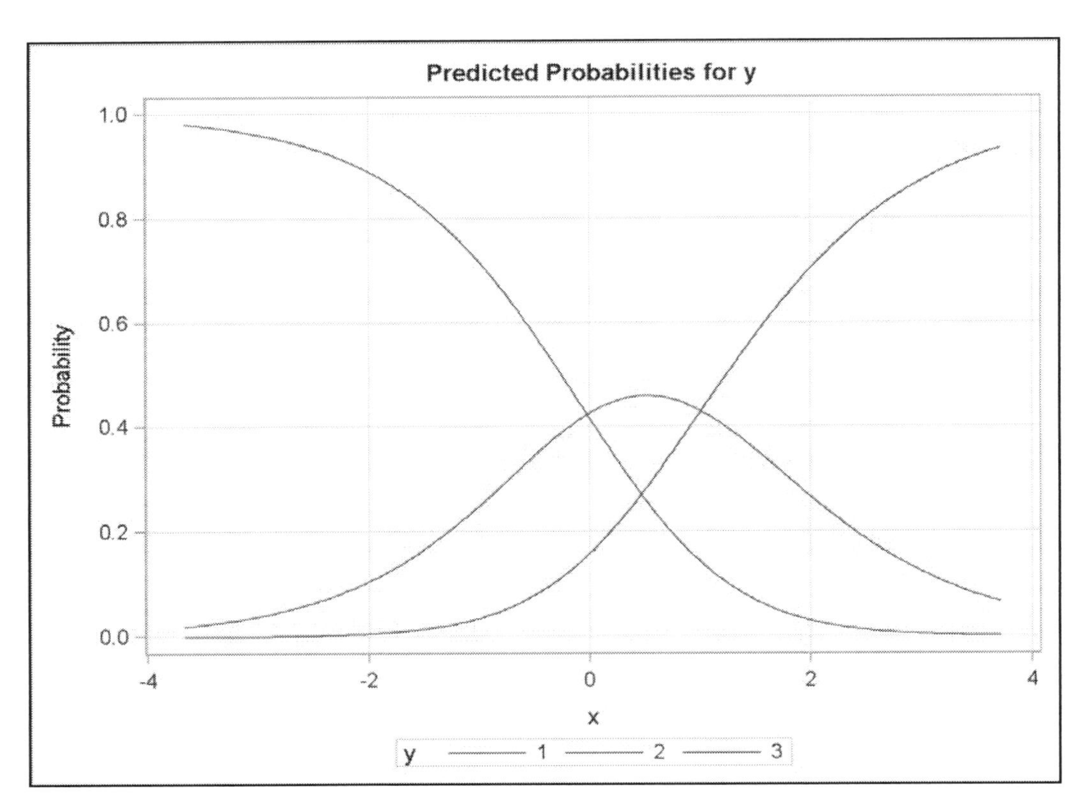

What's going on here? The diagram in Figure 5.1 helps explain this situation. The structure of the model is such that increases in x tend to move individuals from category 2 to category 3. But increases in x also tend to move individuals from category 1 to category 2. When x is very low, most of the cases are in category 1, so most of the movement is from 1 to 2 and the proportion of cases in 2 goes up. Eventually, however, there are few cases left in 1 so most of the movement is from 2 to 3, and the proportion remaining in 2 declines.

Figure 5.1 Movement of Individuals as x Increases

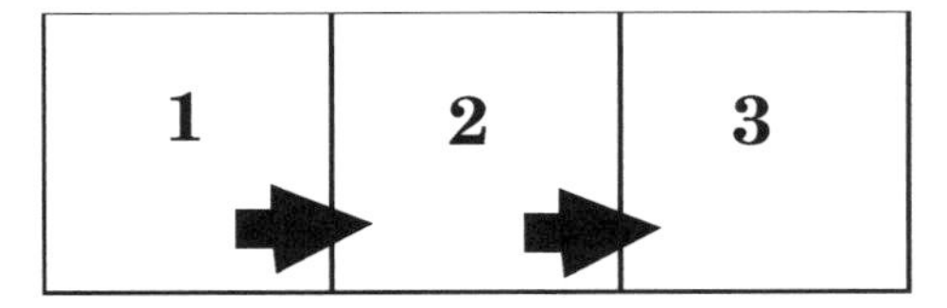

The upshot of this phenomenon is that multinomial logit coefficients must always be interpreted as effects on contrasts between pairs of categories, never on the probability of being in a particular category.

Chapter 6

Logistic Regression for Ordered Categories

6.1 Introduction

In the previous chapter we studied the multinomial logit model for dependent variables with three or more *unordered* categories. When the categories are ordered, it is sometimes useful to ignore the ordering and estimate the unordered multinomial model. However, there are two reasons for preferring models that take the ordering into account:

- They are simpler, and therefore easier to interpret.
- Hypothesis tests are more powerful.

The disadvantage of ordered models is that they impose restrictions that may not be consistent with the data. So whenever you use an ordered model, it's important to test whether its restrictions are valid.

Unlike the unordered case, there are several different ways of generalizing the logit model to handle ordered categories. In this chapter, we'll examine three:

- The cumulative logit model
- The adjacent categories model
- The continuation ratio model

When the dependent variable has only two categories, all three models reduce to the usual binary logit model.

Of the three, the cumulative logit model—also known as the ordered logit or ordinal logit model—is the most widely applicable, and the one that is most easily used in SAS. The adjacent categories model is also an attractive general approach, but SAS can only estimate the model when the data are grouped. The continuation ratio model is more specialized. It's designed for situations in which the ordered categories represent a longitudinal progression through stages.

6.2 Cumulative Logit Model: Example

The cumulative logit model (McCullagh 1980) could hardly be easier to implement in SAS. Whenever PROC LOGISTIC encounters more than two categories on the dependent variable, it automatically estimates a cumulative logit model. Let's try it and see what we get. Then we'll step back and discuss the underlying model.

In Chapter 5 we analyzed data in which the dependent variable was the response to the question, "If you found a wallet on the street, would you (1) keep the wallet and the money, (2) keep the money and return the wallet, or (3) return both the wallet and the money?" Although we treated these response categories as unordered, there's also an obvious ordering: 1 is the most unethical response, 3 is the most ethical, and 2 is in the middle. To fit a cumulative logit model to these data, we submit the program:

```
PROC LOGISTIC DATA=wallet;
  MODEL wallet = male business punish explain;
RUN;
```

(In case you've forgotten, the explanatory variables are described in Section 5.2.) Results are shown in Output 6.1.

The output looks virtually identical to output for a binary logit model, except for something new in the middle: the "Score Test for the Proportional Odds Assumption" table. I'll explain this in more detail later. For now, it's sufficient to say that this statistic tests whether the ordinal restrictions are valid, and high *p*-values are desirable. In this case, we find no reason to reject the model. Another difference in the output is that there are two intercepts instead of just one. In general, the number of intercepts for all the ordinal models in this chapter is one less than the number of categories on the dependent variable. But like the intercept in a linear model, these rarely have any substantive interest.

Output 6.1 PROC LOGISTIC Output for Cumulative Logit Model, Wallet Data

Model Information	
Data Set	MY.WALLET
Response Variable	wallet
Number of Response Levels	3
Model	cumulative logit
Optimization Technique	Fisher's scoring

Number of Observations Read	195
Number of Observations Used	195

Response Profile		
Ordered Value	wallet	Total Frequency
1	1	24
2	2	50
3	3	121

Probabilities modeled are cumulated over the lower Ordered Values.

Model Convergence Status
Convergence criterion (GCONV=1E-8) satisfied.

Score Test for the Proportional Odds Assumption		
Chi-Square	DF	Pr > ChiSq
5.1514	4	0.2721

Model Fit Statistics		
Criterion	Intercept Only	Intercept and Covariates
AIC	356.140	319.367
SC	362.686	339.005
-2 Log L	352.140	307.367

Testing Global Null Hypothesis: BETA=0			
Test	Chi-Square	DF	Pr > ChiSq
Likelihood Ratio	44.7727	4	<.0001
Score	40.8753	4	<.0001
Wald	38.5746	4	<.0001

Analysis of Maximum Likelihood Estimates						
Parameter		DF	Estimate	Standard Error	Wald Chi-Square	Pr > ChiSq
Intercept	1	1	-3.2691	0.5612	33.9325	<.0001
Intercept	2	1	-1.4913	0.5085	8.6012	0.0034
male		1	1.0636	0.3255	10.6771	0.0011
business		1	0.7370	0.3515	4.3973	0.0360
punish		1	0.6874	0.2246	9.3644	0.0022
explain		1	-1.0452	0.3392	9.4972	0.0021

Odds Ratio Estimates			
Effect	Point Estimate	95% Wald Confidence Limits	
male	2.897	1.531	5.483
business	2.090	1.049	4.161
punish	1.989	1.280	3.089
explain	0.352	0.181	0.684

Turning to the more familiar portions of the output, we see from the global tests that there is strong evidence that at least one of the coefficients is not 0. This is further confirmed by the lower portion of the output which shows that all the coefficients have p-values below .05. In the next section, we'll see how the parameter estimates and odds ratios can be interpreted in a fashion that is nearly identical to interpretation in a binary logit model.

Comparing Output 6.1 with the multinomial logit results for the same data (Output 5.1), we find a few major differences. The most obvious is that the multinomial model has two coefficients for every explanatory variable while the cumulative model has only one. This makes the cumulative model much simpler to present and interpret. A less apparent difference is that the *p*-values for the tests of whether each variable has an effect on the dependent variable are all lower in the cumulative model. This is crucial for BUSINESS, which has a *p*-value of .10 in the multinomial model (with the 2 degrees of freedom Type 3 test) but .036 in the cumulative model. The differences in *p*-values illustrate the claim I made earlier, that hypothesis tests in the cumulative logit model are more powerful.

6.3 Cumulative Logit Model: Explanation

I'll explain the cumulative logit model in three different ways. I begin with an intuitive, nonmathematical description. Then I present a formal definition of the model in terms of cumulative probabilities. Finally, I explain how the model can be derived from a latent variable model.

Here's the nonmathematical explanation. Suppose you wanted to analyze the WALLET data, but all you had was a binary logit program. One solution to that dilemma is to group two of the three categories together so that you have a dichotomy. Of course, to respect the ordering of the variable, you would only want to combine categories 1 and 2 or 2 and 3, but not 1 and 3. To get as equal a split as possible on the dependent variable, the best grouping is to combine categories 1 and 2 together and leave 3 by itself. That way, there are 74 cases in the (1, 2) category and 121 cases in category 3. Here's the SAS code for doing that:

```
DATA a;
  SET wallet;
  IF wallet=1 THEN wallet=2;
PROC LOGISTIC DATA=a;
  MODEL wallet = male business punish explain;
RUN;
```

The results are shown in Output 6.2.

Output 6.2 Logistic Results for Dichotomized Wallet Data

Analysis of Maximum Likelihood Estimates					
Parameter	DF	Estimate	Standard Error	Wald Chi-Square	Pr > ChiSq
Intercept	1	-1.3189	0.5465	5.8243	0.0158
male	1	1.1845	0.3408	12.0824	0.0005
business	1	0.6357	0.3812	2.7808	0.0954
punish	1	0.5071	0.2474	4.2030	0.0404
explain	1	-1.0200	0.3662	7.7606	0.0053

There's nothing sacred about that grouping, however. We could also group 2 and 3 together (for 171 cases) and leave 1 by itself (with 24 cases). The alternative code is:

```
DATA b;
  SET wallet;
  IF wallet=3 THEN wallet=2;
PROC LOGISTIC DATA=b;
  MODEL wallet = male business punish explain;
RUN;
```

The results are shown in Output 6.3.

Output 6.3 Results for Alternative Dichotomization of Wallet Data

Analysis of Maximum Likelihood Estimates					
Parameter	DF	Estimate	Standard Error	Wald Chi-Square	Pr > ChiSq
Intercept	1	-3.7153	0.8017	21.4793	<.0001
male	1	0.8268	0.5290	2.4426	0.1181
business	1	1.0129	0.5142	3.8810	0.0488
punish	1	1.0108	0.3075	10.8032	0.0010
explain	1	-1.2760	0.5112	6.2311	0.0126

What does this have to do with the cumulative logit model? Just this: the cumulative model constrains the coefficients in these two binary logit models to be the same but allows the intercepts to differ. Comparing Output 6.2 and Output 6.3, it appears that the coefficients are *not* identical. But the cumulative logit model just chalks the differences up to random error and estimates a sort of weighted average of the corresponding coefficients. Comparing Output 6.1 with Output 6.2 and Output 6.3, we see that in every case the cumulative logit coefficient is in between the two binary coefficients. Moreover, the two intercepts in Output 6.1 are quite close to the intercepts in Output 6.2 and Output 6.3.

In short, the cumulative logit model says that it doesn't make any difference how you dichotomize the dependent variable—the effects of the explanatory variables are always the same. However, as we saw in Output 6.1, we do get a test of whether this is a reasonable constraint. The score test of the proportional odds assumption is a test of the null hypothesis that the corresponding coefficients in Output 6.2 and Output 6.3 are equal. The 4 degrees of freedom correspond to the four pairs of coefficients. With a *p*-value of .27, there's clearly insufficient evidence to reject the null hypothesis. In other words, the differences between Output 6.2 and Output 6.3 could easily be due to chance.

Now let's formalize the model in a way that will apply to an indefinite number of categories on the dependent variable. As in the unordered multinomial logit model, let p_{ij} be the probability that individual i falls into category j of the dependent variable. We assume that the categories are ordered in the sequence $j=1, \ldots, J$. Next we define *cumulative probabilities*

$$F_{ij} = \sum_{m=1}^{j} p_{im}$$

In words, F_{ij} is the probability that individual i is in the jth category *or lower*. Each F_{ij} corresponds to a different dichotomization of the dependent variable. We then specify the model as a set of $J-1$ equations,

$$\log\left(\frac{F_{ij}}{1-F_{ij}}\right) = \alpha_j + \boldsymbol{\beta}\mathbf{x}_i \qquad j = 1,\ldots,J-1 \tag{6.1}$$

where $\boldsymbol{\beta}\mathbf{x}_i = \beta_1 x_{i1} + \ldots + \beta_k x_{ik}$. Although there is a single set of coefficients, there is a different intercept for each of the equations.

The way I have specified the model, the explanatory variables predict the probability of being in a *lower* category rather than in a higher category. That's the default in LOGISTIC, which is indicated in Output 6.1 by the phrase "Probabilities modeled are cumulated over the lower Ordered Values" just after the frequencies in the "Response Profile" table. But we could have just as easily defined the cumulative probabilities as

$$F_{ij} = \sum_{m=j}^{J} p_{im} ,$$

which is the probability of being in the jth category or *higher*. Then the model predicts the probability of being in a *higher* category. That's what you get if you use the DESCENDING option on the MODEL statement (after the dependent variable). If you use this option (which can be abbreviated DESC), the intercepts will be different from those in the default mode, and all the coefficients will change sign.

That's all there is to the model. As usual, we'll estimate it by maximum likelihood, although weighted least squares is a reasonable alternative when the data are grouped.

Here's a third way to think about the cumulative logit model. Like the binary logit model, the cumulative logit model can be viewed as arising from a continuous variable that's been chopped up into categories. Let z_i be a continuous random variable that depends on a set of explanatory variables $\mathbf{x}_i$ according to a linear model:

$$z_i = \alpha^* + \boldsymbol{\beta}^* \mathbf{x}_i + \sigma \varepsilon_i \ . \tag{6.2}$$

We don't observe z directly, however. Instead, there is a set of cut points or *thresholds*, $\tau_1, \ldots, \tau_{J-1}$, that are used to transform z into the observed variable y according to the following rules:

$$\begin{aligned} y &= 1 \text{ if } \tau_1 < z \\ y &= 2 \text{ if } \tau_2 < z \leq \tau_1 \\ &\ \ \vdots \\ y &= J \text{ if } z \leq \tau_{J-1} \end{aligned} \tag{6.3}$$

If we assume that ε_i has a standard logistic distribution, it follows that the dependence of y on $\mathbf{x}$ is given by the cumulative logit model in equation (6.1). The coefficients in equation (6.1) are related to the coefficients in equation (6.2) by

$$\begin{aligned} \alpha_j &= \frac{\alpha^* - \tau_j}{\sigma} \\ \boldsymbol{\beta} &= \boldsymbol{\beta}^* / \sigma. \end{aligned} \tag{6.4}$$

Because the scale parameter σ is not identified, we can't recover the underlying coefficients. But if we test the null hypothesis that an observed coefficient is 0, this is equivalent to testing whether the underlying coefficient is 0.

The remarkable thing about this model is that the $\boldsymbol{\beta}$ coefficients don't depend on the placement of the thresholds. That means that some of the τ's may be close together while others may be far apart, but the effects of the explanatory variables stay the same. The position of the τ's does, of course, affect the intercepts and the relative numbers of cases that fall into the different categories. But there is a sense in which the cumulative logit model makes no assumptions about the distances between observed categories.

In the latent variable interpretation of the *binary* logit model, we saw that if ε had a standard normal distribution, we got a probit model. And if ε had a standard extreme-value distribution, we got a complementary log-log model. The same results generalize to the cumulative model, although I won't discuss these models further. As in the binary case, you can invoke these models in LOGISTIC with the LINK=PROBIT or LINK=CLOGLOG options in the MODEL statement.

6.4 Cumulative Logit Model: Practical Considerations

Now we're ready to interpret the coefficients in Output 6.1. Because I didn't use the DESCENDING option, the model predicts the probability of being in a lower category and, in this case, that means less ethical behavior. Each reported odds ratio can be interpreted as the effect of the variable on the odds of being in a lower rather than in a higher category, *without regard to how you dichotomize the outcome*. For example, the adjusted odds ratio for males is 2.897. We can say, then, that the odds of a male being in a lower category (rather than a higher category) are nearly three times the odds for a female. Among those in the business school, the odds of being in a lower category are a little more than twice the odds for those in other schools. For PUNISH, the adjusted odds ratio of 1.99 tells us that each one-level increase in PUNISH doubles the odds of being in a lower category rather than a higher category. Finally, those students whose parents explained their punishments have odds of being in a lower category that's only about 35% of the odds for those parents who did not explain.

How should the intercepts be interpreted? My preferred answer is, "Don't bother." But if you really must know, the first intercept, –3.2691 in Output 6.1, is the predicted log-odds of being in category 1 rather than in categories 2 or 3, when all the explanatory variables have values of 0. (Actually, it's impossible for PUNISH to be 0 because it only has values of 1, 2, or 3.) Exponentiating gives us .038. So the predicted odds of being in category 1 versus the other two is .038. Similarly, the second intercept, –1.4913, is the log-odds of being in categories 1 or 2 rather than in category 3, when all the *x* variables are 0. Exponentiating gives us odds of .225. So people are more than four times as likely to be in category 3 than in the other two categories if they have values of 0 on all the independent variables.

A word of caution about the score test for the "proportional odds assumption:" the *SAS/STAT User's Guide* warns that the test may tend to reject the null hypothesis more often than is warranted. In my own experience, if there are many independent variables and if the sample size is large, this test will usually produce *p*-values below .05. I don't think that means that the cumulative logit model should always be rejected in such cases. For one thing, changing the specification of the model can often markedly change the *p*-value for this test; for example, adding or deleting variables, adding or deleting interactions, or adding or deleting polynomial terms. If you're really concerned about low *p*-values for this test, you may find it useful to estimate separate models for different dichotomizations of the dependent variable. That way, you can get a better sense of which variables have coefficients that are not invariant across dichotomizations. In any case, you can always go back to the unordered multinomial model.

How many categories can the dependent variable have in a cumulative logit model? Unlike the multinomial logit model, the cumulative model does *not* get more difficult to interpret as you add more categories. On the other hand, each additional category does add one intercept to the model. That means that the model won't work for rank-ordered data in which no two individuals have the same rank—there would be more parameters to estimate than observations in the sample. As a very rough rule of thumb, I would say that it's reasonable to estimate a cumulative logit model if there are at least 10 observations for each category on the dependent variable. Of course, you also need enough observations overall to get decent estimates of the coefficients. As the number of categories on the dependent variable gets larger, ordinary linear regression becomes an attractive alternative, especially if the distribution of cases is not highly skewed.

Many of the issues and options discussed in Chapters 2 and 3 for the binary logit model also apply to the cumulative logit model. For example, everything I said about confidence intervals is true for the cumulative logit model. Multicollinearity is a comparable problem in the cumulative model, and the approach to diagnosis is the same. Problems of non-convergence and infinite parameter estimates can occur in the cumulative model, although with somewhat reduced frequency. That's because, as the number of categories on the dependent variable increases, it's harder to find explanatory variables that perfectly predict the dependent variable. The generalized R^2 is available for the cumulative logit model, and its formula is identical to equation (3.10).

One difference between binary and cumulative models in LOGISTIC is that many regression diagnostics are not available for the cumulative model. For example, you can't produce any of the diagnostic plots discussed in Chapter 3. Using the OUTPUT statement, you can't get any of the residuals or influence statistics. However, you can produce predicted probabilities for each individual and each category, together with upper and lower confidence limits if you wish. And you can plot the predicted values against the explanatory variables with the EFFECTPLOT statement.

There are two methods for getting predicted values for the model in Output 6.1. I'll first explain the traditional method, which is a little tricky. Then I'll describe a newer, more straightforward method. The traditional method is implemented by including the statement

```
OUTPUT OUT=a PRED=p;
```

after the MODEL statement. For each observation in the original data set, there are two records in the OUTPUT data set, one for each of the first two categories on the dependent variable. (With J categories on the dependent variable, there will be J–1 predicted values for each observation.) The first 10 records in data set A (corresponding to the first five cases in the original data set) are shown in Output 6.4. The first two records are for the first person in the original data set. The first record gives a predicted probability of .13076 that WALLET has a value of 1. The second line gives the predicted probability that WALLET will have a value *less than or equal to* 2. To get the predicted probability that WALLET is actually equal to 2, we must subtract the preceding probability, i.e., .47092–.13076=.34016. To get the predicted probability that WALLET=3, we must subtract the predicted probability that WALLET=2 from 1, that is, 1–.47092=.52908. Following this method, the predicted probabilities for the next person in the output (observations 3 and 4) are .05024, .18812, and

.76164 for levels 1, 2, and 3, respectively. Keep in mind that when you use the DESCENDING option, the predicted probabilities are the probabilities that the dependent variable has values *greater than or equal to* the specified level.

Output 6.4 Predicted Values for Wallet Data (First Five Persons)

Obs	wallet	male	business	punish	explain	_LEVEL_	p
1	2	0	0	2	0	1	0.13076
2	2	0	0	2	0	2	0.47091
3	2	0	0	2	1	1	0.05024
4	2	0	0	2	1	2	0.23836
5	3	0	0	1	1	1	0.02591
6	3	0	0	1	1	2	0.13597
7	3	0	0	2	0	1	0.13076
8	3	0	0	2	0	2	0.47091
9	1	1	0	1	1	1	0.07154
10	1	1	0	1	1	2	0.31314

There is also a more straightforward method that we already saw in Chapter 5 for the unordered multinomial model. It is invoked with the statement:

```
OUTPUT OUT=b PREDPROBS=I;
```

This produces a SAS data that contains one record for each record in the original data set. The first five records in the data set are shown in Output 6.5. For each category of the dependent variable, there is a predicted probability that the observation will fall into that category, with variable names IP_1, IP_2, and IP_3.

Output 6.5 Predicted Values for Wallet Data (First Five Persons)

Obs	wallet	male	business	punish	explain	_FROM_	_INTO_	IP_1	IP_2	IP_3
1	2	0	0	2	0	2	3	0.13076	0.34015	0.52909
2	2	0	0	2	1	2	3	0.05024	0.18812	0.76164
3	3	0	0	1	1	3	3	0.02591	0.11006	0.86403
4	3	0	0	2	0	3	3	0.13076	0.34015	0.52909
5	1	1	0	1	1	1	3	0.07154	0.24159	0.68686

6.5 Cumulative Logit Model: Contingency Tables

The cumulative logit model can be very useful in analyzing contingency tables. Consider Table 6.1, which was tabulated by Sloane and Morgan (1996) from the General Social Survey. Our goal is to estimate a model for the dependence of happiness on year and marital status.

Table 6.1 General Happiness by Marital Status and Year

		Very Happy	Pretty Happy	Not Too Happy
1974	Married	473	493	93
	Unmarried	84	231	99
1984	Married	332	387	62
	Unmarried	150	347	117
1994	Married	571	793	112
	Unmarried	257	889	234

Here's the SAS program to read the table:

```
DATA happy;
  INPUT year married happy count;
  DATALINES;
1 1 1 473
1 1 2 493
1 1 3  93
1 0 1  84
1 0 2 231
1 0 3  99
2 1 1 332
2 1 2 387
2 1 3  62
2 0 1 150
2 0 2 347
2 0 3 117
3 1 1 571
3 1 2 793
3 1 3 112
3 0 1 257
3 0 2 889
3 0 3 234
;
```

Notice that I've coded HAPPY so that 1 is very happy and 3 is not too happy.

To fit the cumulative logit model, we run the following program, with results shown in Output 6.6:

```
PROC LOGISTIC DATA=happy;
  FREQ count;
  CLASS year / PARAM=GLM;
  MODEL happy = married year / AGGREGATE SCALE=NONE;
RUN;
```

Output 6.6 Cumulative Logit Results for Happiness Data

Score Test for the Proportional Odds Assumption		
Chi-Square	DF	Pr > ChiSq
26.5204	3	<.0001

Deviance and Pearson Goodness-of-Fit Statistics				
Criterion	Value	DF	Value/DF	Pr > ChiSq
Deviance	30.9709	7	4.4244	<.0001
Pearson	31.4060	7	4.4866	<.0001

Testing Global Null Hypothesis: BETA=0			
Test	Chi-Square	DF	Pr > ChiSq
Likelihood Ratio	356.3435	3	<.0001
Score	348.4470	3	<.0001
Wald	340.2626	3	<.0001

Type 3 Analysis of Effects			
Effect	DF	Wald Chi-Square	Pr > ChiSq
married	1	322.9185	<.0001
year	2	3.7616	0.1525

Analysis of Maximum Likelihood Estimates						
Parameter		DF	Estimate	Standard Error	Wald Chi-Square	Pr > ChiSq
Intercept	1	1	-1.3919	0.0513	735.2275	<.0001
Intercept	2	1	1.4160	0.0517	749.6434	<.0001
married		1	0.9931	0.0553	322.9185	<.0001
year	1	1	0.0716	0.0636	1.2673	0.2603
year	2	1	0.1195	0.0640	3.4827	0.0620
year	3	0	0	.	.	.

Odds Ratio Estimates			
Effect	Point Estimate	95% Wald Confidence Limits	
married	2.700	2.422	3.008
year 1 vs 3	1.074	0.948	1.217
year 2 vs 3	1.127	0.994	1.278

The first thing we see in Output 6.6 is that the score test for the proportional odds assumption indicates fairly decisive rejection of the model. This is corroborated by the overall goodness-of-fit tests (obtained with the AGGREGATE and SCALE=NONE options), both of which have *p*-values less than .0001. The score test has 3 degrees of freedom corresponding to the constraints imposed on the three coefficients in the model. Roughly speaking, the score test (chi-square=26.5) can be thought of as one component of the overall deviance (chi-square=31). So we can say that approximately 85% of the deviance stems from the constraints imposed by the cumulative logit model. The remaining 15% (and 4 degrees of freedom) comes from possible interactions between year and marital status that have been excluded from the model.

Should we reject the model? Keep in mind that the sample size is quite large (5,724 cases), so it may be hard to find any parsimonious model with a *p*-value above .05. But let's postpone a decision until we examine more evidence. Turning to the lower part of the output, we see strong evidence that married people report greater happiness than the unmarried people, but little evidence for change over time. Neither of the individual year coefficients is statistically significant. The Type 3 test that both coefficients are 0 is also nonsignificant. So let's try deleting the year variables and see what happens (Output 6.7).

Output 6.7 Happiness Model with Year Deleted

Score Test for the Proportional Odds Assumption		
Chi-Square	DF	Pr > ChiSq
0.3513	1	0.5534

Deviance and Pearson Goodness-of-Fit Statistics				
Criterion	Value	DF	Value/DF	Pr > ChiSq
Deviance	0.3508	1	0.3508	0.5537
Pearson	0.3513	1	0.3513	0.5534

Number of unique profiles: 2

Analysis of Maximum Likelihood Estimates						
Parameter		DF	Estimate	Standard Error	Wald Chi-Square	Pr > ChiSq
Intercept	1	1	-1.3497	0.0459	865.8068	<.0001
Intercept	2	1	1.4569	0.0468	968.1472	<.0001
married		1	1.0017	0.0545	337.2287	<.0001

Odds Ratio Estimates			
Effect	Point Estimate	95% Wald Confidence Limits	
married	2.723	2.447	3.030

Now the model fits great. Of course, there's only 1 degree of freedom in the score test because only one coefficient is constrained across the two implicit equations. The deviance and Pearson chi-squares also have only 1 degree of freedom because LOGISTIC has regrouped the data after the elimination of the year variables. Interpreting the marital status effect, we find that married people have odds of higher happiness that are nearly three times the odds for unmarried people.

It's tempting to leave it at this, but the fact that the score statistic declined so dramatically with the deletion of the year variables suggests that something else is going on. To see what it might be, let's fit separate models for the two ways of dichotomizing the happiness variable.

```
DATA a;
 SET happy;
 lesshap=happy GE 2;
 nottoo=happy EQ 3;
PROC LOGISTIC DATA=a;
  FREQ count;
  CLASS year / PARAM=GLM;
  MODEL lesshap=married year;
PROC LOGISTIC DATA=a;
  FREQ count;
  CLASS year / PARAM=GLM;
  MODEL nottoo=married year;
RUN;
```

Results are shown in Output 6.8. If the cumulative logit model is correct, the coefficients for the three variables should be the same in the two models. That's nearly true for MARRIED. But the two YEAR variables have negative coefficients in the first model and positive coefficients in the second. Three of the four coefficients are significant at beyond the .01 level. This is surely the cause of the difficulty with the cumulative logit model. What seems to be happening is that 1994 is different from the other two years, but not in a way that could be described as a uniform increase or decrease in happiness. Rather, in that year there were relatively more cases in the middle category and fewer in the two extreme categories.

Output 6.8 Results from Alternative Dichotomizations of General Happiness

```
1 vs. (2,3)
```

Analysis of Maximum Likelihood Estimates						
Parameter		DF	Estimate	Standard Error	Wald Chi-Square	Pr > ChiSq
Intercept		1	-1.4615	0.0574	647.4056	<.0001
married		1	0.9931	0.0624	253.2451	<.0001
year	1	1	0.2204	0.0699	9.9348	0.0016
year	2	1	0.2271	0.0716	10.0650	0.0015
year	3	0	0	.	.	.

```
(1,2) vs. 3
```

Analysis of Maximum Likelihood Estimates						
Parameter		DF	Estimate	Standard Error	Wald Chi-Square	Pr > ChiSq
Intercept		1	1.5549	0.0644	583.3901	<.0001
married		1	1.0109	0.0842	144.1691	<.0001
year	1	1	-0.3037	0.0996	9.3082	0.0023
year	2	1	-0.1109	0.0999	1.2331	0.2668
year	3	0	0	.	.	.

It's possible to generalize the cumulative logit model to accommodate patterns like this. Specifically, one can model the parameter σ in equation (6.2) as a function of explanatory variables. Specifically, we can write

$$\log \sigma_i = \boldsymbol{\gamma} \mathbf{z}_i \tag{6.5}$$

where $\mathbf{z}_i$ is a vector of predictor variables that may or may not be the same as $\mathbf{x}_i$. Although this model cannot be estimated with PROC LOGISTIC, it's an option with PROC QLIM. Here's the code:

```
DATA happy2;
  set happy;
  yr94=year EQ 3;
  yr84=year EQ 2;
PROC QLIM data=happy2;
  FREQ count;
  MODEL happy=married / DISCRETE(D=LOGISTIC);
  HETERO happy ~ yr94 yr84 / NOCONST;
RUN;
```

The DATA step is necessary because QLIM, although it supports the CLASS statement, ran into technical difficulties using YEAR as a CLASS variable in the HETERO statement. So, instead, I created dummy variables for two of the three values of YEAR. The DISCRETE option on the MODEL statement says to treat HAPPY as a discrete variable with a logit link function—the default is a probit link. The HETERO statement requests that the residual variance σ^2 be modeled as a function of YR94 and YR84. The NOCONST option suppresses the constant in the log-linear equation for σ. That produces the most common version of the model.

Output 6.9 Results from QLIM with Heterogeneous Disturbance

Parameter Estimates					
Parameter	DF	Estimate	Standard Error	t Value	Approx Pr > |t|
Intercept	1	1.224813	0.055617	22.02	<.0001
married	1	-0.904331	0.056926	-15.89	<.0001
_Limit2	1	2.549680	0.085444	29.84	<.0001
_H.yr94	1	-0.335531	0.080389	-4.17	<.0001
_H.yr84	1	-0.081647	0.094153	-0.87	0.3858

Results are consistent with what we have already seen. The sign of the MARRIED coefficient is the opposite of what we got with PROC LOGISTIC, because QLIM predicts the probability of being in a higher category, not a lower category, of the dependent variable. The two parameters preceded by _H are effects on the residual variance. We see that the variance in 1994 is significantly lower than in 1974, indicating that in 1994 people were more likely to be in the middle category and less likely to be in the extremes. The parameter labeled _LIMIT2 is the functional equivalent of the second intercept in the PROC LOGISTIC model.

6.6 Adjacent Categories Model

Another general model for ordered categorical data is the adjacent categories model. As before, we let p_{ij} be the probability that individual i falls into category j of the dependent variable, and we assume that the categories are ordered in the sequence j=1, …, J. Now take any pair of categories that are adjacent, such as j and j+1. We can write a logit model for the contrast between these two categories as a function of explanatory variables:

$$\log\left(\frac{p_{i,j+1}}{p_{ij}}\right)=\alpha_j+\boldsymbol{\beta}_j\mathbf{x}_i \qquad j=1,\ldots,J-1, \tag{6.6}$$

where $\boldsymbol{\beta}_j\mathbf{x}_i=\beta_{j1}x_{i1}+\ldots+\beta_{jk}x_{ik}$. There are J–1 of these paired contrasts. It turns out that this is just another way of writing the multinomial logit model for unordered categories. In other words, equation (6.6) is equivalent to equation (5.1). To get the adjacent categories model for ordered data, we impose a constraint on this set of equations. Specifically, we assume that $\boldsymbol{\beta}_j=\boldsymbol{\beta}$ for all j. In other words, instead of having a different set of coefficients for every adjacent pair, there is only one set for the lot of them. So our adjacent categories model becomes

$$\log\left(\frac{p_{i,j+1}}{p_{ij}}\right)=\alpha_j+\boldsymbol{\beta}\mathbf{x}_i \qquad j=1,\ldots,J-1 \tag{6.7}$$

Notice that the right-hand side is identical to equation (6.1), which defines the cumulative model, but the left-hand side compares individual categories rather than grouped, cumulative categories.

Although the adjacent categories model is a special case of the multinomial logit model, PROC LOGISTIC doesn't allow the imposition of the appropriate constraints. The model can be estimated with PROC CATMOD if the data can be grouped into a contingency table. However, CATMOD uses the method of weighted least squares rather than maximum likelihood. Here's how to do it for the happiness data that we analyzed in the previous section:

```
PROC CATMOD DATA=happy;
  WEIGHT count;
  RESPONSE ALOGIT;
  MODEL happy = _RESPONSE_ married year / PARAM=REF;
RUN;
```

CATMOD doesn't have a FREQ statement, but the WEIGHT statement performs the same function. The RESPONSE statement with the ALOGIT option invokes the adjacent categories function for the dependent variable, as in equation (6.6). Putting _RESPONSE_ in the MODEL statement tells CATMOD to estimate a single set of coefficients, as in equation (6.7), rather than a different set for each pair of categories. By default CATMOD treats all predictor variables as CLASS variables. But like LOGISTIC, CATMOD also codes those variables by using "effect parameterization." The PARAM=REF option overrides that default to produce indicator variable coding. Results are shown in Output 6.10.

The first thing to notice is that the residual chi-square test indicates that the model doesn't fit. In fact, the value is fairly close to the deviance and Pearson chi-squares we got for the cumulative logit model in Output 6.6. As in that output, we also find a strong effect of marital status but little evidence for an effect of calendar year. Unlike LOGISTIC, CATMOD parameterizes the model to predict higher rather than lower values of the response variable. But because MARRIED is treated as a CLASS variable, the reference category is 1 rather than 0. So the coefficient for MARRIED has the same sign as in Output 6.6, and the corresponding odds ratio is exp(.8035)=2.23. This tells us that whenever we compare adjacent happiness categories, married people have more than double the odds of being in the happier category compared with unmarried people.

Output 6.10 CATMOD Output for Adjacent Categories Model, Happiness Data

Analysis of Variance			
Source	**DF**	**Chi-Square**	**Pr > ChiSq**
Intercept	1	1097.28	<.0001
RESPONSE	1	1323.91	<.0001
married	1	314.22	<.0001
year	2	3.07	0.2159
Residual	7	35.74	<.0001

Analysis of Weighted Least Squares Estimates					
Parameter		**Estimate**	**Standard Error**	**Chi-Square**	**Pr > ChiSq**
Intercept		-1.8856	0.0569	1097.28	<.0001
RESPONSE	**1**	2.1474	0.0590	1323.91	<.0001
married	**0**	0.8035	0.0453	314.22	<.0001
year	**1**	-0.0353	0.0516	0.47	0.4945
	2	-0.0916	0.0523	3.06	0.0801

Because the model doesn't fit well, we can do what we did with the cumulative logit model: fit separate equations for each of the two adjacent pairs. In this case, we can easily accomplish that by removing _RESPONSE_ from the MODEL statement. The results are shown in Output 6.11.

Output 6.11 Adjacent Categories Model with Two Sets of Coefficients

Analysis of Variance			
Source	**DF**	**Chi-Square**	**Pr > ChiSq**
Intercept	2	606.88	<.0001
married	2	315.39	<.0001
year	4	30.66	<.0001
Residual	4	4.16	0.3845

Analysis of Weighted Least Squares Estimates						
Parameter		Function Number	Estimate	Standard Error	Chi-Square	Pr > ChiSq
Intercept		1	0.3492	0.0497	49.45	<.0001
		2	-2.0122	0.0818	604.75	<.0001
married	0	1	0.8619	0.0646	178.10	<.0001
	0	2	0.7087	0.0863	67.46	<.0001
year	1	1	-0.2900	0.0726	15.94	<.0001
	1	2	0.4036	0.1024	15.53	<.0001
	2	1	-0.2606	0.0734	12.60	0.0004
	2	2	0.1932	0.1028	3.53	0.0603

These results are quite similar to what is shown in Output 6.8 for alternative dichotomizations. The MARRIED coefficient is quite stable across the two equations, as is the coefficient for YEAR 1 (1992). But the YEAR 2 (1994) coefficient is much larger in the first equation than in the second equation. Again, the message is that in 1994 the middle category of the response variable was more strongly favored over the two extremes.

Despite clear-cut differences in the formulation and interpretation of the cumulative logit model and the adjacent categories model, the two models tend to yield very similar conclusions in practice. I generally prefer the cumulative model, both for its appealing latent variable interpretation and for its ready availability in software. But the adjacent categories model has one advantage, at least in principle: it's easy to formulate models with selective constraints on coefficients (although such models can't be estimated in CATMOD). For example, we could force the MARRIED coefficient to be the same for all category pairs, but allow the year coefficients to be different. In fact, we'll see how to do that in Chapter 10, when we estimate the adjacent categories model by using the equivalent loglinear model.

6.7 Continuation Ratio Model

The cumulative logit and adjacent categories models are reasonable candidates for almost any ordered categorical variable. Our third model—the continuation ratio model—is more specialized. It's most appropriate when the ordered categories represent a progression through stages, so that individuals must pass through each lower stage before they go on to higher stages. In those situations, the continuation ratio model is more attractive than the other two.

Because the model is most often applied to educational attainment, let's use that as our example here. Table 6.2 gives frequency counts for three categories of educational attainment for 2,294 males who failed to pass the Armed Forces Qualification Test, cross classified by race, age, and father's education (Fienberg 2007). The fact that they failed to pass the test is irrelevant to our purposes.

Table 6.2 Educational Attainment by Race, Age, and Father's Education

			Educational Attainment		
Race	Age	Father's Education*	Grammar School	Some High School	High School Graduate
White	<22	1	39	29	8
		2	4	8	1
		3	11	9	6
		4	48	17	8
	≥22	1	231	115	51
		2	17	21	13
		3	18	28	45
		4	197	111	35
Black	<22	1	19	40	19
		2	5	17	7
		3	2	14	3
	≥22	4	49	79	24
		1	110	133	103
		2	18	38	25
		3	11	25	18
		4	178	206	81

*1=grammar school, 2=some high school, 3=high school graduate, 4=not available.

We could, of course, estimate a cumulative logit or an adjacent categories model for this table. We could also ignore the ordinality altogether and estimate a multinomial logit model. But let's think about how the attainment process works in this case. The first step is to get past grammar school. We could estimate a binary logit model with the dependent variable equal to 1 if the person went past grammar school and 0 if he didn't, essentially collapsing the two higher categories. The next step is to graduate from high school, *given that you started high school*. For this step, we could eliminate all the men who never started high school (those in the first column of frequency counts) and estimate a logit model with the dependent variable equal to 1 if the person graduated and 0 if he didn't.

Now, suppose we believe that the effects of the explanatory variables are the same at each step. It would make sense, then, to constrain the coefficients to be equal across steps so that we only get a single set of coefficients. That's the essential idea of the continuation ratio model.

Let's formalize the model for J categories on the dependent variable. As usual we assume that the categories are ordered in the sequence j=1, …, J. Let y_i be the dependent variable that can take on one of these J values. Define A_{ij} as the probability that individual i advances to stage j+1, given that he made it to stage j. More formally,

$$A_{ij} = \Pr(y_i > j \mid y_i \geq j).$$

We then specify J–1 logit equations,

$$\log\left(\frac{A_{ij}}{1-A_{ij}}\right) = \alpha_j + \boldsymbol{\beta}\mathbf{x}_i \qquad j = 1,\ldots, J-1 \tag{6.8}$$

where $\boldsymbol{\beta}\mathbf{x}_i = \beta_1 x_{i1} + \ldots + \beta_k x_{ik}$. Again, the right-hand side of equation (6.8) is just like the cumulative logit model and the adjacent categories model, with a separate intercept for each stage but a single set of coefficients. The left-hand side can also be rewritten in terms of the original probabilities, p_{ij}:

$$\log\left(\frac{A_{ij}}{1-A_{ij}}\right) = \log\left[\frac{\sum_{m=j+1}^{J} p_{im}}{p_{ij}}\right]$$

How can this model be estimated? Surprisingly, we can do it with an ordinary binary logit procedure. The trick is in the construction of the data set. Again, imagine doing a sequence of binary logit analyses. For each stage, you construct a data set that excludes all individuals who didn't make it to that stage, with a dummy dependent variable indicating whether or not the individual advanced to the next stage. Now, instead of doing separate analyses, we combine these data sets into a single set, including a variable indicating which stage the data came from. Finally, we estimate a single binary logit model on the combined data.

Here's how to do it for the data in Table 6.2. First, we read in the data as a contingency table, with each record corresponding to one of the cells. (To save space, two records are placed on each data line, which is accommodated by the @@ on the INPUT statement).

```
DATA afqt;
  INPUT white old faed ed count @@;
  DATALINES;
1  0  1  1   39   0  0  1  1   19
1  0  1  2   29   0  0  1  2   40
1  0  1  3    8   0  0  1  3   19
1  0  2  1    4   0  0  2  1    5
1  0  2  2    8   0  0  2  2   17
1  0  2  3    1   0  0  2  3    7
1  0  3  1   11   0  0  3  1    2
1  0  3  2    9   0  0  3  2   14
1  0  3  3    6   0  0  3  3    3
1  0  4  1   48   0  0  4  1   49
1  0  4  2   17   0  0  4  2   79
1  0  4  3    8   0  0  4  3   24
1  1  1  1  231   0  1  1  1  110
1  1  1  2  115   0  1  1  2  133
1  1  1  3   51   0  1  1  3  103
1  1  2  1   17   0  1  2  1   18
1  1  2  2   21   0  1  2  2   38
1  1  2  3   13   0  1  2  3   25
1  1  3  1   18   0  1  3  1   11
1  1  3  2   28   0  1  3  2   25
1  1  3  3   45   0  1  3  3   18
1  1  4  1  197   0  1  4  1  178
1  1  4  2  111   0  1  4  2  206
1  1  4  3   35   0  1  4  3   81
;
```

The data set for the first stage is then constructed as follows:

```
DATA first;
  SET afqt;
  stage=1;
  advance = ed GE 2;
RUN;
```

For the second stage, we have the following:

```
DATA second;
  SET afqt;
  stage=2;
  IF ed=1 THEN DELETE;
  advance = ed EQ 3;
RUN;
```

The two data sets are concatenated into a single set as follows:

```
DATA concat;
  SET first second;
RUN;
```

Alternatively, we can create the combined data set in a single DATA step:

```
DATA combined;
  SET afqt;
   stage=1;
   advance = ed GE 2;
  OUTPUT;
   stage=2;
   IF ed=1 THEN DELETE;
   advance = ed EQ 3;
  OUTPUT;
RUN;
```

Now we're ready to estimate the model with PROC LOGISTIC:

```
PROC LOGISTIC DATA=combined;
  FREQ count;
  CLASS faed / PARAM=REF;
  MODEL advance(EVENT='1')=stage white old faed / AGGREGATE
    SCALE=NONE;
RUN;
```

The results in Output 6.12 show strong effects of race, father's education, and stage, but little or no effect of age (at the time of the AFQT examination). White men are only about half as likely to advance to the next stage as black men. As you might expect, with increasing education of the father, the odds of advancing go up substantially, with the lowest probability of advancement among men whose father's education was "unavailable."

Output 6.12 PROC LOGISTIC Results for Continuation Ratio Model

Deviance and Pearson Goodness-of-Fit Statistics				
Criterion	**Value**	**DF**	**Value/DF**	**Pr > ChiSq**
Deviance	85.1090	25	3.4044	<.0001
Pearson	84.6686	25	3.3867	<.0001

Analysis of Maximum Likelihood Estimates						
Parameter		DF	Estimate	Standard Error	Wald Chi-Square	Pr > ChiSq
Intercept		1	1.5268	0.1352	127.4323	<.0001
stage		1	-1.1785	0.0753	244.9951	<.0001
white		1	-0.6918	0.0731	89.5732	<.0001
old		1	0.0919	0.0867	1.1228	0.2893
faed	1	1	0.2678	0.0776	11.9123	0.0006
faed	2	1	0.6771	0.1336	25.7059	<.0001
faed	3	1	1.2031	0.1339	80.7279	<.0001

Odds Ratio Estimates			
Effect	Point Estimate	95% Wald Confidence Limits	
stage	0.308	0.266	0.357
white	0.501	0.434	0.578
old	1.096	0.925	1.299
faed 1 vs 4	1.307	1.123	1.522
faed 2 vs 4	1.968	1.515	2.557
faed 3 vs 4	3.330	2.562	4.330

The highly significant deviance and Pearson chi-square statistics suggest that something about this model is not right. Like the other ordinal models, this one imposes some restrictions, namely that the effects of the explanatory variables are the same at each stage. We can test this by including interactions between STAGE and the other three variables:

```
MODEL advance(EVENT='1')=stage white old faed stage*white
   stage*old stage*faed;
```

This produces the Type 3 statistics shown in Output 6.13. Clearly there are interactions with all three variables, which means that each variable has different effects on advancement into high school and advancement to high school graduation.

Output 6.13 Tests for Interaction between STAGE and Other Variables

Type 3 Analysis of Effects			
Effect	DF	Wald Chi-Square	Pr > ChiSq
stage	1	123.8551	<.0001
white	1	60.3596	<.0001
old	1	6.0710	0.0137
faed	3	27.4223	<.0001
stage*white	1	24.3852	<.0001
stage*old	1	8.6909	0.0032
stage*faed	3	11.0070	0.0117

When none of the variables has an effect that is invariant across stages, it's probably better to estimate separate models for each stage rather than attempting to interpret the interactions in the combined data set. We can do that with the following program:

```
PROC LOGISTIC DATA=first;
  FREQ count;
  CLASS faed /PARAM=GLM;
  MODEL advance(EVENT='1')= white old faed;
PROC LOGISTIC DATA=second;
  FREQ count;
  CLASS faed /PARAM=GLM;
  MODEL advance(EVENT='1')= white old faed;
RUN;
```

The results in Output 6.14 show markedly different effects for age and race at the two stages. Whites were substantially less likely than blacks to advance from grade school to high school, but only slightly less likely to advance from high school entrance to graduation. The opposite pattern is found for age with no effect at the earlier stage, but older males are about 50% more likely than younger ones to advance to high school graduation. The father's education has strong positive effects at both stages, but the ordinal pattern is not as clear cut at the second stage.

Output 6.14 Separate Logit Models for Two Levels of Advancement

Advancement from Grammar School to High School

Analysis of Maximum Likelihood Estimates						
Parameter		DF	Estimate	Standard Error	Wald Chi-Square	Pr > ChiSq
Intercept		1	0.6424	0.1129	32.3763	<.0001
white		1	-0.9388	0.0897	109.5316	<.0001
old		1	-0.1092	0.1110	0.9680	0.3252
faed	1	1	0.1781	0.0952	3.4960	0.0615
faed	2	1	0.9198	0.1895	23.5634	<.0001
faed	3	1	1.3376	0.1910	49.0229	<.0001
faed	4	0	0	.	.	.

Advancement from High School Entrance to High School Graduation

Analysis of Maximum Likelihood Estimates						
Parameter		DF	Estimate	Standard Error	Wald Chi-Square	Pr > ChiSq
Intercept		1	-1.3217	0.1570	70.8575	<.0001
white		1	-0.1811	0.1245	2.1166	0.1457
old		1	0.4427	0.1508	8.6214	0.0033
faed	1	1	0.4721	0.1348	12.2748	0.0005
faed	2	1	0.4425	0.2079	4.5303	0.0333
faed	3	1	1.0273	0.1949	27.7924	<.0001
faed	4	0	0	.	.	.

For these data, then, the constrained continuation ratio model is a failure, its attractive parsimony apparently spurious. Nevertheless, even when a single equation model is inappropriate, there can be conceptual advantages in viewing the process in terms of stages. Output 6.14 is probably more meaningfully interpreted than a multinomial logit model with an equal number of coefficients. Another potential advantage of the continuation ratio approach—one not applicable to the data in Table 6.2—is the possibility of stage-dependent explanatory variables. Variables like parental income, number of persons in the household, and presence of the father are likely to vary over time, and this variation can be easily incorporated into the modeling process. In constructing the working data set, you simply redefine the variables to have different values at each of the stages.

There are several other points worth noting about the continuation ratio method. First, you may have observed that in the combined data set, the same people may appear more than once. For the education example, everyone who had some high school contributed two observations to the combined data set. Usually this raises red flags because of possible dependence among the multiple observations. While there isn't space here to go into the details, there is absolutely no problem of dependence. For each individual, the probability of a particular outcome is factored into a set of conditional probabilities that have the same structure as independent observations.

Second, the continuation ratio approach is closely related to discrete-time methods for survival analysis. In that setting, the aim is to model the length of time until some event occurs, and time is measured in discrete units like years. Each year is treated as a stage and "advancement" means getting to the next year without experiencing the event. For further details, see Chapter 7 of my book, *Survival Analysis Using SAS: A Practical Guide*, Second Edition (2010).

Third, just as there are cumulative probit and cumulative complementary log-log models, one can easily estimate continuation ratio models using the probit or complementary log-log functions. The complementary log-log model is particularly attractive for event history applications because it is the discrete-time equivalent of the proportional hazards model used in Cox regression. I'll close this chapter with a statistical curiosity: the complementary log-log continuation ratio model is mathematically equivalent to the cumulative complementary log-log model (McCullagh 1980).

Chapter 7

Discrete Choice Analysis

7.1 Introduction

In economics and marketing research, discrete choice analysis has become a popular approach to the study of consumer choice. Although the method is not well known in sociology or political science, it ought to be. There are many potential applications of this methodology that have been overlooked because of ignorance of the available tools.

In discrete choice problems, people choose from among a set of options that are available to them—political candidates for example—and the aim is to discover what variables affect their choices. While this may sound like a job for the multinomial logit model of Chapter 5, it differs in two respects:

- The explanatory variables can include characteristics of the choice options as well as variables describing the relationship between the chooser and the option.
- The set of available options can vary across individuals in the analysis.

Discrete choice analysis is usually based on the *conditional logit model*, a name that—unfortunately—has also been used for quite different models. This model is very easily estimated using the STRATA statement in PROC LOGISTIC. But I'll also show how to estimate more general models using PROC PHREG and PROC MDC.

7.2 Chocolate Example

The discrete choice model has often been used in studies of consumer preference. Consider the following experiment (SAS Institute 1995). Ten people were each given eight chocolate bars with varying characteristics. After eating the bars, they were asked to choose the one they liked best. The bars were distinguished by the following variables:

DARK 0=milk chocolate, 1=dark chocolate

SOFT 0=hard center, 1=soft center

NUTS 0=no nuts, 1=nuts

Each subject tasted eight chocolate bars with all possible combinations of these characteristics. The aim was to determine which characteristics of the bars affected people's choices. Here are the data:

```
DATA chocs;
   INPUT id choose dark soft nuts @@;
   DATALINES;
 1 0 0 0 0    1 0 0 0 1    1 0 0 1 0    1 0 0 1 1
 1 1 1 0 0    1 0 1 0 1    1 0 1 1 0    1 0 1 1 1
 2 0 0 0 0    2 0 0 0 1    2 0 0 1 0    2 0 0 1 1
 2 0 1 0 0    2 1 1 0 1    2 0 1 1 0    2 0 1 1 1
 3 0 0 0 0    3 0 0 0 1    3 0 0 1 0    3 0 0 1 1
 3 0 1 0 0    3 0 1 0 1    3 1 1 1 0    3 0 1 1 1
 4 0 0 0 0    4 0 0 0 1    4 0 0 1 0    4 0 0 1 1
 4 1 1 0 0    4 0 1 0 1    4 0 1 1 0    4 0 1 1 1
 5 0 0 0 0    5 1 0 0 1    5 0 0 1 0    5 0 0 1 1
 5 0 1 0 0    5 0 1 0 1    5 0 1 1 0    5 0 1 1 1
 6 0 0 0 0    6 0 0 0 1    6 0 0 1 0    6 0 0 1 1
 6 0 1 0 0    6 1 1 0 1    6 0 1 1 0    6 0 1 1 1
 7 0 0 0 0    7 1 0 0 1    7 0 0 1 0    7 0 0 1 1
 7 0 1 0 0    7 0 1 0 1    7 0 1 1 0    7 0 1 1 1
 8 0 0 0 0    8 0 0 0 1    8 0 0 1 0    8 0 0 1 1
 8 0 1 0 0    8 1 1 0 1    8 0 1 1 0    8 0 1 1 1
 9 0 0 0 0    9 0 0 0 1    9 0 0 1 0    9 0 0 1 1
 9 0 1 0 0    9 1 1 0 1    9 0 1 1 0    9 0 1 1 1
```

```
10 0 0 0 0   10 0 0 0 1  10 0 0 1 0   10 0 0 1 1
10 0 1 0 0   10 1 1 0 1  10 0 1 1 0   10 0 1 1 1
;
```

There are eight observations for each person. (To save space, each line contains four observations, with the @@ operator indicating that there are multiple observations per line.) The first variable, ID, is a unique identifier for each person so that we can keep track of who is making the choices. The second variable, CHOOSE, has a value of 1 if the person chose that particular chocolate bar, otherwise 0. So for every set of eight observations, only one of them has CHOOSE=1 while the other seven have CHOOSE=0.

We'll first see how to estimate the conditional logit model for these data using PROC LOGISTIC. Then, in the next section, we'll examine some of the mathematics underlying this model. Here is the LOGISTIC code:

```
PROC LOGISTIC DATA=chocs;
  MODEL choose(EVENT='1')=dark soft nuts;
  STRATA id;
RUN;
```

The STRATA statement tells LOGISTIC how to group the observations together for each person. The results are shown in Output 7.1.

Output 7.1 PROC LOGISTIC Output for Discrete Choice Model, Chocolate Data

Conditional Analysis

Model Information	
Data Set	WORK.CHOCS
Response Variable	choose
Number of Response Levels	2
Number of Strata	10
Model	binary logit
Optimization Technique	Newton-Raphson ridge

Number of Observations Read	80
Number of Observations Used	80

Response Profile		
Ordered Value	choose	Total Frequency
1	0	70
2	1	10

Probability modeled is choose=1.

Strata Summary				
Response Pattern	choose		Number of Strata	Frequency
	0	1		
1	7	1	10	80

Newton-Raphson Ridge Optimization
Without Parameter Scaling

Convergence criterion (GCONV=1E-8) satisfied.

Model Fit Statistics		
Criterion	Without Covariates	With Covariates
AIC	41.589	34.727
SC	41.589	41.873
-2 Log L	41.589	28.727

Testing Global Null Hypothesis: BETA=0			
Test	Chi-Square	DF	Pr > ChiSq
Likelihood Ratio	12.8618	3	0.0049
Score	11.6000	3	0.0089
Wald	8.9275	3	0.0303

Analysis of Maximum Likelihood Estimates					
Parameter	DF	Estimate	Standard Error	Wald Chi-Square	Pr > ChiSq
dark	1	1.3863	0.7906	3.0749	0.0795
soft	1	-2.1972	1.0541	4.3450	0.0371
nuts	1	0.8473	0.6901	1.5076	0.2195

Odds Ratio Estimates			
Effect	Point Estimate	95% Wald Confidence Limits	
dark	4.000	0.849	18.836
soft	0.111	0.014	0.877
nuts	2.333	0.603	9.023

Output 7.1 looks much like previous LOGISTIC output, but there are a couple of additions. The second line describes this as a "conditional analysis." Then, a little later, there is a new table titled "Strata Summary." We see from this table that there are 10 strata, one for each person. Each stratum has exactly one record with CHOOSE=1, and seven records with CHOOSE=0. Later, we'll see an example in which there are several different response patterns. There are also a couple of things that are missing from the output. No intercept is reported, which is a characteristic of conditional logistic regression that I'll explain in the next section. And we don't get the usual "Association of Predicted Probabilities and Observed Responses" table.

We do get the usual global tests of the null hypothesis that all the coefficients are 0. In this case, all three are statistically significant, although the *p*-value for the Wald test is quite a bit higher than for the other two. As in other models, the likelihood-ratio test is generally preferred.

In the next table, we find coefficients and associated statistics. Only the SOFT coefficient has a *p*-value below .05. However, I was suspicious of those *p*-values for two reasons: (a) the sample size is on the small side with only 80 records contributed from 10 persons, and (b) there is a marked divergence between the global likelihood ratio chi-square and the Wald chi-square. To resolve this uncertainty, I did exact tests by adding the following statement to the previous program:

```
EXACT dark soft nuts / ESTIMATES=BOTH;
```

This produced the following two-sided p-values for the coefficients and odds ratios, essentially confirming the large-sample approximations in Output 7.1.

	p-value
DARK	.109
SOFT	.021
NUTS	.343

As we'll see in the next section, the coefficients can be interpreted as effects on the log odds of choosing a bar of that particular kind relative to other kinds. As usual, these coefficients are exponentiated to produce odds ratio estimates. We see, for example, that the odds of choosing a dark chocolate bar are four times the odds of choosing a milk chocolate bar. The odds of choosing a soft bar are only about one-ninth the odds of choosing a hard bar. People were twice as likely to choose bars with nuts, but that coefficient was far from significant.

Notice that this data set does not contain any variables describing the persons in the study. Even if variables like gender, education, or race were available in the data, they could not be included as variables in the model. Because a person's gender is the same for all eight chocolate bars, gender cannot help predict why a person chooses one bar rather than another. On the other hand, it would be possible to include interactions between gender and the other variables in the model. For example, we could test whether a soft center is more important for men than for women. We'll see an example of such interaction terms in Section 7.4.

7.3 Model and Estimation

Now that we've seen how to do a discrete choice analysis, let's take a closer look at the model being estimated. Suppose we have i=1, … , n individuals and each individual is presented with j=1, … , J_i options. We write the number of possible choices as J_i to indicate that different individuals may have different sets of options. That wasn't true for the chocolate example, but it's an important feature of the model that distinguishes it from the multinomial logit model of Chapter 5. Let $y_{ij}=1$ if individual i chooses option j, otherwise 0, and let $\mathbf{x}_{ij}$ be a vector of explanatory variables describing option j for person i. This set of explanatory variables may include dummy variables for the various options as well as interactions between option characteristics and individual characteristics. However, it does *not* include an intercept term.

The *conditional logit model* introduced by McFadden (1974) is

$$\Pr(y_{ij}=1)=\frac{e^{\boldsymbol{\beta}\mathbf{x}_{ij}}}{e^{\boldsymbol{\beta}\mathbf{x}_{i1}}+e^{\boldsymbol{\beta}\mathbf{x}_{i2}}+\ldots+e^{\boldsymbol{\beta}\mathbf{x}_{iJ_i}}}. \tag{7.1}$$

This equation is quite similar to equation (5.2) for the multinomial logit model. In that model, however, there was one **x** vector for each individual and separate coefficient vectors $\boldsymbol{\beta}_j$ for each of the possible outcomes. For the conditional logit model, there is only one coefficient vector but different **x** vectors for each outcome. Despite these apparent differences, the standard multinomial logit model can be shown to be a special case of the conditional logit model by appropriate coding of the explanatory variables, as illustrated below in Output 7.6 and 7.7.

From equation (7.1) we can see why there is no intercept term in the conditional logit model. Suppose we insisted on including an intercept by writing each of the terms in the numerator and denominator as $e^{\beta_0+\boldsymbol{\beta}\mathbf{x}_{ij}}$. This, of course, is equal to $e^{\beta_0}e^{\boldsymbol{\beta}\mathbf{x}_{ij}}$. But because e^{β_0} appears in every term, this quantity cancels out of the fraction.

Equation (7.1) implies that the logit for comparing any two options *j* and *k* is given by

$$\log\left(\frac{\Pr(y_{ij}=1)}{\Pr(y_{ik}=1)}\right)=\boldsymbol{\beta}(\mathbf{x}_{ij}-\mathbf{x}_{ik}).$$

Equivalently, the odds that person *i* will choose *j* over *k* is given by

$$\exp\{\boldsymbol{\beta}(\mathbf{x}_{ij}-\mathbf{x}_{ik})\}.$$

We estimate the model by maximum likelihood. The likelihood function is the product of *n* factors, each having the form of equation (7.1) for the particular option that is chosen.

Is this a reasonable model for individual choice? One justification is that the model can be derived from a latent variable model called the random utility model. Suppose that each individual *i* has a stable preference for option *j*, denoted by μ_{ij}. But suppose the actual *utility* U_{ij} for a particular option varies randomly around μ_{ij} so that

$$U_{ij} = \mu_{ij}+\varepsilon_{ij} \tag{7.2}$$

where ε_{ij} is a random variable having a standard extreme value distribution with constant variance. Assume further that the ε_{ij}'s are independent across the different options. If people choose the option with the highest utility U_{ij}, and if the logarithm of μ_{ij} is a linear function of the explanatory variables, it follows that the probability that person *i* chooses option *j* is given by equation (7.1).

If these assumptions are credible, then the conditional logit model is reasonable. There has been some concern, however, about the assumption of independence of the random variables ε_j. This assumption (together with the extreme-value distributional assumption) implies a condition that is known as *independence of irrelevant alternatives*, frequently denoted IIA (read i-i-a, not 2a). This condition means that the odds of choosing option *j* rather than option *k* are not affected by what other options are available.

It's not hard to come up with counter examples to the IIA assumption. A classic example is the blue-bus/red-bus dilemma. Suppose I have two ways of getting to work, by car or by blue bus. My probability of taking the car is .8, which gives a car/blue bus odds of .8/.2=4. Now suppose another bus is introduced, identical in every way to the blue bus except that it's red, so I now have three options. Because I'm indifferent to bus color, it's reasonable to suppose that my probability of taking the car will still be .8 and that the remaining .2 will be equally divided between the two colors of buses. But then the conditional odds of choosing a car rather than a blue bus are .8/.1=8, a violation of the IIA condition.

The IIA assumption can only be empirically tested when some sample members have different choice sets. That suggests that when everyone in the sample is presented with the same choice set, the IIA assumption may not be a serious problem. Here's a simple model that illustrates that claim. Suppose that the potential choice set contains only three alternatives, labeled 1, 2, and 3. When people are presented with alternatives 1 and 2 only, their relative odds are given by $\log(p_1 / p_2) = \alpha + \boldsymbol{\beta}\mathbf{x}$, where $\mathbf{x}$ is a vector of explanatory variables. When people are presented with alternatives 1, 2, *and* 3, the relative odds of 1 versus 2 are given by $\log(p_1 / p_2) = \alpha + \mu + \boldsymbol{\beta}\mathbf{x}$. Clearly, this violates the IIA condition because of the additional μ term. But as long as everyone in the sample is presented with the same set of alternatives, conventional logit analysis will produce consistent and efficient estimates of $\boldsymbol{\beta}$.

Efforts to relax the IIA assumption have focused on introducing correlations among the ε_j's in the random utility model, allowing them to be heteroscedastic and allowing for alternative distributional shapes. In the last section of this chapter, we'll take a look at some of these more advanced discrete choice methods which are available in PROC MDC.

7.4 Travel Example

Now let's look at a more complicated application of the conditional logit model. In a survey of transportation usage between three Australian cities, 210 people were asked which of four options they used on their most recent trip: air, train, bus, or car (Hensher and Bradley 1993). They were also asked about certain characteristics of all four travel options, regardless of whether they used them. As in the previous example, a separate observation is constructed for each option for each person, giving a total of 840 observations. The variables are

ID	A unique identifier for each person
MODE	1=air, 2=train, 3=bus, 4=car
CHOICE	1=chose that mode, 0=didn't choose
TTME	Estimated terminal waiting time
COST	Estimated total cost for all stages
TIME	Estimated total time in vehicle for all stages
HINC	Household income in thousands
PSIZE	Traveling party size

Output 7.2 displays the data for the first 20 observations (five persons). For each person, the last two variables are the same across all four options. The preceding three variables—TTME, COST, and TIME—describe relationships between the person and the option. For each person, they vary over options; and for each option, they vary over persons.

Output 7.2 First 20 Observations in Travel Data Set

Obs	ID	MODE	TTME	HINC	PSIZE	CHOICE	TIME	COST
1	1	1	69	35	1	0	100	59
2	1	2	34	35	1	0	372	31
3	1	3	35	35	1	0	417	25
4	1	4	0	35	1	1	180	10
5	2	1	64	30	2	0	68	58
6	2	2	44	30	2	0	354	31
7	2	3	53	30	2	0	399	25
8	2	4	0	30	2	1	255	11
9	3	1	69	40	1	0	125	115
10	3	2	34	40	1	0	892	98
11	3	3	35	40	1	0	882	53
12	3	4	0	40	1	1	720	23

Obs	ID	MODE	TTME	HINC	PSIZE	CHOICE	TIME	COST
13	4	1	64	70	3	0	68	49
14	4	2	44	70	3	0	354	26
15	4	3	53	70	3	0	399	21
16	4	4	0	70	3	1	180	5
17	5	1	64	45	2	0	144	60
18	5	2	44	45	2	0	404	32
19	5	3	53	45	2	0	449	26
20	5	4	0	45	2	1	600	8

Let's begin with a simple model that includes only those variables that vary over options:

```
PROC LOGISTIC DATA=travel;
  MODEL choice(EVENT='1') = ttme time cost;
  STRATA id;
RUN;
```

Results in Output 7.3 show highly significant negative effects of terminal time and travel time. Each additional minute of terminal waiting time reduces the odds of choosing that alternative by 3.3%, calculated as (100×(1–.967)). Each additional minute of travel time reduces the odds by about 0.2%. The COST coefficient doesn't quite reach statistical significance, and it has the "wrong" sign—greater cost appears to increase the odds that an alternative is chosen.

Output 7.3 Conditional PROC LOGISTIC Model for the Determinants of Travel Mode

Probability modeled is CHOICE=1.

Strata Summary				
Response Pattern	CHOICE 0	CHOICE 1	Number of Strata	Frequency
1	3	1	210	840

Testing Global Null Hypothesis: BETA=0			
Test	Chi-Square	DF	Pr > ChiSq
Likelihood Ratio	88.5263	3	<.0001
Score	86.0063	3	<.0001
Wald	74.9939	3	<.0001

Analysis of Maximum Likelihood Estimates					
Parameter	DF	Estimate	Standard Error	Wald Chi-Square	Pr > ChiSq
TTME	1	-0.0340	0.00464	53.5600	<.0001
TIME	1	-0.00219	0.000458	22.9151	<.0001
COST	1	0.00889	0.00488	3.3239	0.0683

Odds Ratio Estimates			
Effect	Point Estimate	95% Wald Confidence Limits	
TTME	0.967	0.958	0.975
TIME	0.998	0.997	0.999
COST	1.009	0.999	1.019

Now let's estimate a model that excludes the variables in the previous output, but includes MODE as a CLASS variable:

```
PROC LOGISTIC DATA=travel;
  CLASS mode / PARAM=GLM;
  MODEL choice(EVENT='1') = mode;
  STRATA id;
RUN;
```

The coefficients in Output 7.4 represent the relative preferences for each mode compared with a car, the omitted mode. Keeping in mind that MODE has values of 1 for air, 2 for train, and 3 for bus, we see that bus travel is significantly less attractive than car travel—the odds of choosing bus travel are only about half the odds of choosing car travel. But there are no appreciable differences in the preferences for air, train, or car.

Output 7.4 Travel Model with MODE as a CLASS Variable

Type 3 Analysis of Effects			
Effect	DF	Wald Chi-Square	Pr > ChiSq
MODE	3	12.6107	0.0056

Analysis of Maximum Likelihood Estimates						
Parameter		DF	Estimate	Standard Error	Wald Chi-Square	Pr > ChiSq
MODE	1	1	-0.0171	0.1849	0.0085	0.9263
MODE	2	1	0.0656	0.1812	0.1311	0.7173
MODE	3	1	-0.6763	0.2242	9.0973	0.0026
MODE	4	0	0	.	.	.

Odds Ratio Estimates			
Effect	Point Estimate	95% Wald Confidence Limits	
MODE 1 vs 4	0.983	0.684	1.412
MODE 2 vs 4	1.068	0.749	1.523
MODE 3 vs 4	0.508	0.328	0.789

In Output 7.5, we see the results of combining these two models. The COST coefficient is now in the expected direction with a *p*-value below .05. The terminal time coefficient has increased markedly so that now each additional minute results in a 9% reduction in the odds of choice. The coefficients for the mode dummies are all much larger and highly significant. People overwhelmingly prefer *not* to go by car.

What's going on here? The coefficients for the mode dummies in Output 7.5 represent relative preferences for these modes *after* differences in the other variables are taken into account. Because terminal time is 0 for car travel, there's a strong correlation between terminal time and the mode dummies. It appears that the main reason people like to travel by car is the absence of terminal waiting. After terminal time is taken into account, car travel seems to be unappealing. The coefficients for the time and cost variables can be interpreted as representing the effects of these variables after taking account of the overall preference for each mode compared with the others.

Output 7.5 Model Combining MODE and Mode Characteristics

Analysis of Maximum Likelihood Estimates						
Parameter		DF	Estimate	Standard Error	Wald Chi-Square	Pr > ChiSq
MODE	1	1	4.7399	0.8675	29.8512	<.0001
MODE	2	1	3.9532	0.4686	71.1828	<.0001
MODE	3	1	3.3062	0.4583	52.0365	<.0001
MODE	4	0	0	.	.	.
TTME		1	-0.0969	0.0103	87.7646	<.0001
TIME		1	-0.00399	0.000849	22.1308	<.0001
COST		1	-0.0139	0.00665	4.3746	0.0365

Odds Ratio Estimates			
Effect	Point Estimate	95% Wald Confidence Limits	
MODE 1 vs 4	114.419	20.896	626.530
MODE 2 vs 4	52.102	20.798	130.523
MODE 3 vs 4	27.282	11.111	66.990
TTME	0.908	0.889	0.926
TIME	0.996	0.994	0.998
COST	0.986	0.973	0.999

The next model removes the three time and cost variables and puts in the interactions between MODE and the two variables that are constant over the four options—income and party size:

```
PROC LOGISTIC DATA=travel;
  CLASS mode / PARAM=REF;
  MODEL choice(EVENT='1') = mode mode*hinc mode*psize;
  STRATA id;
RUN;
```

Results are shown in Output 7.6.

Output 7.6 Conditional Logit Model with Interactions

Type 3 Analysis of Effects			
Effect	DF	Wald Chi-Square	Pr > ChiSq
MODE	3	23.2464	<.0001
HINC*MODE	3	31.9808	<.0001
PSIZE*MODE	3	14.0730	0.0028

Analysis of Maximum Likelihood Estimates						
Parameter		DF	Estimate	Standard Error	Wald Chi-Square	Pr > ChiSq
MODE	1	1	0.9435	0.5498	2.9444	0.0862
MODE	2	1	2.4938	0.5357	21.6702	<.0001
MODE	3	1	1.9780	0.6717	8.6710	0.0032
HINC*MODE	1	1	0.00354	0.0103	0.1183	0.7309
HINC*MODE	2	1	-0.0573	0.0118	23.4211	<.0001
HINC*MODE	3	1	-0.0303	0.0132	5.2597	0.0218
PSIZE*MODE	1	1	-0.6006	0.1992	9.0892	0.0026
PSIZE*MODE	2	1	-0.3098	0.1956	2.5098	0.1131
PSIZE*MODE	3	1	-0.9404	0.3245	8.4011	0.0038

This model illustrates the fact that the standard multinomial logit model discussed in Chapter 5 is a special case of the conditional logit model. The results in Output 7.6 are *identical* to what you get with the following LOGISTIC program, which deletes all records for modes that were not chosen (leaving 210 records) and uses MODE as the dependent variable:

```
PROC LOGISTIC DATA=travel;
  WHERE choice=1;
  MODEL mode(EVENT='1')=hinc psize / LINK=GLOGIT;
RUN;
```

The results from this program are shown in Output 7.7. The intercepts in the multinomial logit output correspond to the coefficients of the three mode dummies in the conditional logit output. The main effects of HINC and PSIZE in the multinomial output correspond to interactions of these variables with the mode dummies in the conditional output.

What do Output 7.6 and Output 7.7 tell us? The coefficients for the mode dummies (the intercepts in Output 7.7) are not particularly informative because they represent the relative log-odds of the four modes when both income and party size are 0, which is a condition that is clearly impossible. We see a highly significant negative effect of the train (mode 2) × income interaction, indicating that with increasing income there is a decreased preference for trains over cars. Output 7.6 doesn't report the odds ratios because the parameters are interactions. (You could get them by using the ODDSRATIO statement described in Chapter 4.) From Output 7.7, however, we see that each thousand-dollar increase in income reduces the odds of choosing a train over a car by 5.6%. A similar effect is found for income on bus preference, though not as large. Increasing party size also reduces the preferences for the public modes over the automobile, although the effect is not quite as significant for trains. For buses, each one-person increase in party size reduces the odds of choosing a bus over a car by over 60%. A plausible explanation is that costs increase directly with party size for the public modes but not for the automobile.

Output 7.7 Multinomial Logit Model

Type 3 Analysis of Effects

Effect	DF	Wald Chi-Square	Pr > ChiSq
HINC	3	31.9765	<.0001
PSIZE	3	14.0706	0.0028

Analysis of Maximum Likelihood Estimates

Parameter	MODE	DF	Estimate	Standard Error	Wald Chi-Square	Pr > ChiSq
Intercept	1	1	0.9435	0.5498	2.9443	0.0862
Intercept	2	1	2.4937	0.5357	21.6689	<.0001
Intercept	3	1	1.9776	0.6717	8.6688	0.0032
HINC	1	1	0.00354	0.0103	0.1183	0.7309
HINC	2	1	-0.0573	0.0118	23.4169	<.0001
HINC	3	1	-0.0303	0.0132	5.2596	0.0218
PSIZE	1	1	-0.6006	0.1992	9.0888	0.0026
PSIZE	2	1	-0.3098	0.1956	2.5102	0.1131
PSIZE	3	1	-0.9401	0.3244	8.3989	0.0038

Odds Ratio Estimates				
		Point	95% Wald	
Effect	MODE	Estimate	Confidence Limits	
HINC	1	1.004	0.983	1.024
HINC	2	0.944	0.923	0.966
HINC	3	0.970	0.945	0.996
PSIZE	1	0.549	0.371	0.810
PSIZE	2	0.734	0.500	1.076
PSIZE	3	0.391	0.207	0.738

The next model (with results shown in Output 7.8) tests this explanation by including the cost and time measures:

```
PROC LOGISTIC DATA=travel;
  CLASS mode / PARAM=REF;
  MODEL choice(EVENT='1') = mode mode*hinc mode*psize ttme
     time cost;
  STRATA id;
RUN;
```

Greene (1992) refers to this sort of model as a *mixed* model because it contains variables that are constant over outcomes and variables that vary over outcomes. (But the term mixed model is more commonly used to describe a very different kind of model for longitudinal and clustered data that is described in Chapter 8.) The coefficients for the time and cost measures are not much different from those in Output 7.5. On the other hand, the effects of income and party size on the odds of choosing a bus over a car have declined dramatically, to the point where they are no longer statistically significant.

Output 7.8 A Mixed Model for Travel Choice

Analysis of Maximum Likelihood Estimates						
Parameter		DF	Estimate	Standard Error	Wald Chi-Square	Pr > ChiSq
MODE	1	1	6.0352	1.1382	28.1158	<.0001
MODE	2	1	5.5735	0.7113	61.3995	<.0001
MODE	3	1	4.5047	0.7958	32.0426	<.0001
HINC*MODE	1	1	0.00748	0.0132	0.3211	0.5710
HINC*MODE	2	1	-0.0592	0.0149	15.8168	<.0001
HINC*MODE	3	1	-0.0209	0.0164	1.6337	0.2012

Analysis of Maximum Likelihood Estimates						
Parameter		DF	Estimate	Standard Error	Wald Chi-Square	Pr > ChiSq
PSIZE*MODE	1	1	-0.9224	0.2585	12.7325	0.0004
PSIZE*MODE	2	1	0.2163	0.2336	0.8569	0.3546
PSIZE*MODE	3	1	-0.1479	0.3428	0.1862	0.6661
TTME		1	-0.1012	0.0111	82.4588	<.0001
TIME		1	-0.00413	0.000893	21.4054	<.0001
COST		1	-0.00867	0.00788	1.2117	0.2710

7.5 Other Applications

The two discrete choice examples that we examined in the previous sections had two features in common:

- Each person had the same set of options.
- Each person could choose one and only one option.

Neither of these features is essential to the discrete choice methodology. In the chocolate bar example, instead of receiving all eight chocolate bars, each person could have been given two chocolate bars and asked to decide which was better. Or some people could get two chocolate bars while others get three or four chocolate bars. In studies of travel mode choice, some cities might not have train service, which removes that option from the choice set of persons traveling to those cities. This creates no special difficulty for data processing and analysis. Each person has a distinct record for every available option—some people may have more records, others may have fewer. The only limitation is that if you want to include dummy variables for each option there has to be a substantial amount of overlap in people's option sets.

If the principal aim is to estimate the effects of the characteristics of the options on each person's choice, there's no need for the option sets to have any items in common. For example, one well-known study examined the factors affecting students' choices of where to attend graduate school (Punj and Staelin 1978). Each person's option set consisted of the schools to which he or she had been accepted. Obviously, this set varied enormously in size and composition. Explanatory variables included such things as student body size, tuition, financial aid, academic reputation, and distance from the applicant. While there was undoubtedly some overlap in students' option sets, the analysis could have proceeded even if every student had been accepted by a completely distinct set of colleges.

The methodology can also be extended to studies in which people are allowed to choose more than one option from among their option sets. Imagine a multi-city study of people's preference for radio stations. Each person's option set consists of the stations broadcasting in his or her city, and the question asked is, "Which radio stations do you listen to regularly?" Obviously, the number of choices will vary greatly from one person to another. Some discrete-choice software can't handle multiple choices, for example, PROC MDC which we'll cover in Section 7.7. But PROC LOGISTIC does it easily with the syntax we've already been using. The only difference is that the choice variable will have the event value ('1' in our examples) for more than one item in the choice set.

7.6 Ranked Data

The discrete-choice approach can be extended even further. Instead of asking people to *choose* one or more items within an option set, we can ask them to *rank* the items on any criterion we specify. Here's an example. In Chapter 2, we studied 147 murder cases that went to a penalty jury to decide between a life sentence and a death sentence. In an effort to get an unbiased measure of the seriousness of the crimes, an auxiliary study was conducted in which 50 trial judges evaluated the murder cases. Each judge was asked to read documents describing 14 or 15 cases in some detail. Then they ranked the cases from 1 to 14 (or 15) in order of assessed culpability with 1 indicating the most serious. Each case was evaluated by four to six judges. There were 10 distinct groups of cases such that (a) every case in a group was evaluated by the same judges and (b) judges evaluating cases in a group saw none of the cases in the other groups. This group structure had no impact on the analysis, however.

Our goal is to estimate a model in which characteristics of the murder cases predict the judges' rankings. The model we'll use—sometimes called the *exploded logit model* (Punj and Staelin 1978, Allison and Christakis 1994)—can be motivated by supposing that judges construct their rankings by making a sequence of choices. They first choose the most serious offense from among the 15 in front of them and assign it a value of 1. They make this choice based on the conditional logit model:

$$\Pr(y_{ij}=1) = \frac{e^{\boldsymbol{\beta}\mathbf{x}_j}}{e^{\boldsymbol{\beta}\mathbf{x}_1}+e^{\boldsymbol{\beta}\mathbf{x}_2}+\ldots+e^{\boldsymbol{\beta}\mathbf{x}_{15}}}\,. \tag{7.3}$$

where y_{ij} is the probability that judge i chooses case j, and $\mathbf{x}_j$ is a vector of characteristics describing case j. After removing the chosen case from the option set, each judge then chooses the most serious case from among the remaining 14. Again, the choice is governed by the conditional logit model, except that now there are only 14 terms in the denominator. This process continues until the last choice is made from among the two that are left. The likelihood for each judge's set of rankings is formed as the product of all the terms that are like the one in equation (7.3). And the overall likelihood is just the product of all the judges' likelihoods. Although the model can be derived in other ways that do not assume sequential choices, this is certainly the most intuitive way to think about it.

How can we estimate this model? It turns out that the likelihood function I just described is identical to the likelihood function for a stratified Cox regression analysis. So we can use PROC PHREG (which does Cox regression) with the dependent variable consisting of ranks (1 through 15). For the penalty trial data, I constructed a data set in which each of 736 records corresponded to a ranking of one murder case by one judge. The data included characteristics of each case and an ID number for the judge. To do the analysis I used the following program:

```
PROC PHREG DATA=judgernk NOSUMMARY;
  MODEL rank=blackd whitvic death / TIES=DISCRETE;
  STRATA judgid;
RUN;
```

The NOSUMMARY option suppresses a large quantity of output that is irrelevant to this application. If every judge had given unique ranks to all cases, the TIES=DISCRETE option would be unnecessary. But in quite a few cases, a judge assigned tied ranks to several of the cases being ranked. Fortunately, PHREG is one of the few Cox regression programs that can handle tied ranks in an appropriate manner. Results are shown in Output 7.9. The column labeled "Hazard Ratio" should actually be interpreted as an odds ratio.

Output 7.9 PROC PHREG Model for Rankings of Murder Cases

Analysis of Maximum Likelihood Estimates						
Parameter	DF	Parameter Estimate	Standard Error	Chi-Square	Pr > ChiSq	Hazard Ratio
BLACKD	1	0.03082	0.09481	0.1057	0.7451	1.031
WHITVIC	1	0.23404	0.10246	5.2180	0.0224	1.264
DEATH	1	0.42688	0.09199	21.5327	<.0001	1.532

Recall that BLACKD is a dummy variable indicating whether the defendant was black, WHITVIC is a dummy variable indicating whether the victim was white, and DEATH is a dummy variable indicating whether the defendant received a death sentence. The case descriptions presented to the judges contained no direct information on any of these three variables. Still, it's not surprising that those cases that got a death sentence were ranked as more serious by the judges. Both juries and judges were responding to the same characteristics of the cases. The hazard ratio for DEATH has the following interpretation. At any point in the choice sequence, cases with death sentences had odds of being chosen (as most serious) that were 53% higher than the odds for cases with life sentences.

It *is* a bit surprising that WHITVIC has a statistically significant coefficient. Cases with white victims had 26% greater odds of being chosen as more serious than cases with a nonwhite victim. This could be explained by either (a) judges somehow guessing the race of the victim from other case characteristics or (b) victim race being associated with other case characteristics that affected judges' rankings.

I also estimated a model with an additional variable CULP. This variable measured severity based on predicted values from a logistic regression of the death penalty cases with DEATH as the dependent variable and numerous case characteristics as independent variables. As shown in Output 7.10, this is a better predictor of the judges' rankings than a death sentence itself. Each increase in the 5-point CULP scale increases the odds of being chosen as more serious by about 30%. Controlling for CULP, cases that received a death sentence are actually *less* likely to be chosen as more serious, although the coefficient is not statistically significant. The WHITVIC effect is virtually unchanged.

Output 7.10 PROC PHREG Ranking Model with Additional Variable

Analysis of Maximum Likelihood Estimates						
Parameter	DF	Parameter Estimate	Standard Error	Chi-Square	Pr > ChiSq	Hazard Ratio
BLACKD	1	0.11951	0.09713	1.5140	0.2185	1.127
WHITVIC	1	0.23698	0.10458	5.1346	0.0235	1.267
DEATH	1	-0.18181	0.13766	1.7443	0.1866	0.834
CULP	1	0.25858	0.04233	37.3245	<.0001	1.295

This approach to the analysis of ranked data can also be used with partial rankings. For example, a ranker may be asked to choose the three best items out of a list of 10 and give them ranks of 1, 2, and 3. If all of the rest are coded as 4, the problem is equivalent to tied ranks, which we already discussed. The PHREG syntax needs no modification.

7.7 More Advanced Models with PROC MDC

Another SAS procedure that can estimate conditional logit models for discrete choice is PROC MDC (for multinomial discrete choice). Compared to PROC LOGISTIC, PROC MDC has a couple of disadvantages: people are not allowed to choose more than one item, and MDC doesn't report odds ratios. But the upside is that MDC can estimate several more advanced models that relax some of the assumptions of the conditional logit model. This section provides only a brief introduction to these models and is not intended to be a thorough treatment.

Let's first see how to duplicate the models for the travel data in Section 7.4. Although PROC MDC in SAS 9.3 supports the CLASS statement, the construction of the design variables and the calculation of the *p*-values is very different than what we have previously seen in PROC LOGISTIC. To avoid difficulties in interpretation, instead of using the CLASS statement, I created three indicator variables to represent the different modes of travel.

```
DATA travel2;
  SET travel;
  air = mode EQ 1;
  train = mode EQ 2;
  bus = mode EQ 3;
RUN;
```

The reference category is travel by car (MODE = 4). Now here's the MDC code for the model that produced Output 7.5, which included the MODE dummies and the mode/person characteristics:

```
PROC MDC DATA=travel2;
  MODEL choice = air train bus ttme time cost / CHOICE=(mode)
    TYPE=clogit;
  ID id;
RUN;
```

The ID statement plays the same role as the STRATA statement in PROC LOGISTIC. TYPE=CLOGIT specifies the conditional logit model. We'll try other models in a moment. The CHOICE option specifies the variable containing integer values that distinguish different choices. For the basic conditional logit model, it's not obvious why this option is needed. But it plays an important role in the more complex models. The main portion of the output is shown in Output 7.11.

Output 7.11 A Conditional Logit Model with PROC MDC

Parameter Estimates					
Parameter	**DF**	**Estimate**	**Standard Error**	**t Value**	**Approx Pr > |t|**
air	1	4.7399	0.8675	5.46	<.0001
train	1	3.9532	0.4686	8.44	<.0001
bus	1	3.3062	0.4583	7.21	<.0001
TTME	1	-0.0969	0.0103	-9.37	<.0001
TIME	1	-0.003995	0.000849	-4.70	<.0001
COST	1	-0.0139	0.006651	-2.09	0.0365

Results in Output 7.11 are essentially the same as those in Output 7.5 produced by PROC LOGISTIC, except that *t*-statistics rather than chi-squares are reported.

Now let's try relaxing some of the assumptions of the conditional logit model. We saw earlier that the conditional logit model can be derived by assuming that choices arise from underlying random utilities, as expressed in equation (7.2). One assumption of that model was that the random variables ε_{ij} all had the same variance. Now let's allow the variances to differ for each choice option j. This is sometimes called the heteroscedastic extreme value (HEV) model because the ε_{ij}'s are assumed to have standard extreme value distributions but different variances (Bhat 1995). Unlike the conditional logit model, the HEV model does not imply the IIA assumption.

Here's a program for estimating the HEV model with the travel data:

```
PROC MDC DATA=travel;
  MODEL choice = ttme time cost / CHOICE=(mode)
    TYPE=HEV;
  ID id;
RUN;
```

Note that I have excluded the three mode indicators from the model. The more advanced models available in MDC often have difficulty estimating models that include dummy variables for the choice options. In fact, even the model without the mode dummies did not converge. To get it to work, I had to change the default numerical integration method by specifying:

```
  MODEL choice = ttme time cost / CHOICE=(mode)
    TYPE=HEV HEV=(INTEGRATE=HARDY);
```

Results are shown in Output 7.12. Coefficients and *p*-values for the three predictor variables are very similar to those for the conditional logit model reported in Output 7.3. So allowing for heteroscedasticity didn't have much impact in this case. The three SCALE parameters are estimates of the variances of ε_{ij} for each mode choice j. To achieve identification, the variance for j=1 is fixed at 1.0.

Despite the fact that the model is reported to have converged, the large variance estimate for SCALE4, along with its much larger standard error, suggests that the algorithm may not be converging properly for these data. Casting further doubt is the "Maximum Absolute Gradient," which is reported to be 180.056. For a model that has properly converged, this should be a number very close to 0.

Output 7.12 A Heteroscedastic Extreme Value Model with PROC MDC

Model Fit Summary	
Dependent Variable	CHOICE
Number of Observations	210
Number of Cases	840
Log Likelihood	-225.35378
Maximum Absolute Gradient	180.05603

Model Fit Summary	
Number of Iterations	21
Optimization Method	Dual Quasi-Newton
AIC	462.70755
Schwarz Criterion	482.79020

Parameter Estimates					
Parameter	DF	Estimate	Standard Error	t Value	Approx Pr > |t|
TTME	1	-0.0415	0.006142	-6.76	<.0001
TIME	1	-0.002087	0.000366	-5.70	<.0001
COST	1	0.009352	0.004613	2.03	0.0426
SCALE2	1	0.7447	0.1262	5.90	<.0001
SCALE3	1	1.0734	0.2004	5.36	<.0001
SCALE4	1	60.8379	602.0463	0.10	0.9195

The conditional logit model is a special case of the HEV model so, in principle, we can do a likelihood ratio test to compare them. The log likelihood for the conditional logit model in Output 7.3 is -246.86. For the HEV model it's -225.35. Taking twice the positive difference between the log-likelihoods yields a chi-square statistic of 43.0. With 3 degrees of freedom (for the three extra SCALE parameters), this statistic has a very low *p*-value. So the HEV model clearly fits significantly better than the conditional logit model. But due to the questionable convergence status of the HEV model, we might still prefer the conditional logit model.

A somewhat more complicated model is the multinomial probit. This model not only allows for heteroscedasticity among the error terms for the different choice options, but also for correlations among them. In addition, instead of assuming that the ε_{ij}'s have an extreme value distribution, we assume that they have a multivariate normal distribution. To achieve identification, the program automatically sets the variances for *two* of the choice options equal to 1. And for one of those two choice options, the correlations with all the other options are set to 0. Here's the code for estimating the model for the travel data:

```
PROC MDC DATA=travel2;
  MODEL choice = ttme time cost / CHOICE=(mode)
    TYPE=MPROBIT;
  ID id;
RUN;
```

Results are shown in Output 7.13. The *p*-values for TTME and TIME are about the same as those for the conditional logit and the HEV models, but the *p*-value for COST is quite a bit higher. Because the coefficients are based on a probit model, they should not be directly compared those in the conditional logit or HEV models, and they cannot be exponentiated to produce odds ratios. STD_1 and STD_2 are the estimated standard deviations of the errors for choice 1 and choice 2, with those for 3 and 4 set to 1. The *p*-values for these estimates show that they are significantly different from 0, but what we really want to know is if they are significantly different from 1, the standard deviation for choices 3 and 4. The 95% confidence intervals for these estimates do include 1, suggesting that we do not have evidence for heteroscedasticity.

The correlation between the error for choice 4 and those for the other choices has been set to 0. The correlation between the error for choice 1 and choice 2 is estimated to be -.96, but it's still not significantly different from 0. The other two correlations were converging toward a number less than -1. But since a correlation less than -1 is not possible, they were restricted at that boundary value. The next two lines for RESTRICT1 and RESTRICT2 give tests for whether the data are consistent with restricting those correlations to -1. The high *p*-values suggest that there is no problem with these restrictions, although it's not clear how correlations of -1 should be interpreted in this case.

Output 7.13 A Mutlivariate Probit Model with PROC MDC

Parameter Estimates						
Parameter	DF	Estimate	Standard Error	t Value	Approx Pr > |t|	Parameter Label
TTME	1	-0.0382	0.006760	-5.65	<.0001	
TIME	1	-0.002727	0.000917	-2.97	0.0029	
COST	1	0.001812	0.004861	0.37	0.7094	
STD_1	1	1.6492	0.6638	2.48	0.0130	
STD_2	1	2.0939	0.6691	3.13	0.0018	
RHO_21	1	-0.9648	0.7882	-1.22	0.2209	
RHO_31	0	-1.0000	0			
RHO_32	0	-1.0000	0			
Restrict1	1	0.5921	1.7909	0.33	0.7418*	Lower BC RHO_31
Restrict2	1	1.4626	1.1732	1.25	0.2133*	Lower BC RHO_32

Because of the normality assumption, the HEV and conditional logit models are not special cases of the multinomial probit model. However, we can compare these models with the BIC (Schwartz) criterion reported in the output. Here are the figures: conditional logit 509.8, HEV 482.8, and multivariate probit 496.5. So the best fitting model (with the lowest BIC) is the HEV model, although the convergence status of that model is questionable.

Now let's consider one other model in MDC that allows for departures from the IIA assumption, the nested logit model. This model presumes that choice options can be nested into groups, with random utilities that are correlated within groups and uncorrelated between groups. For the travel data, for example, we might suppose that all the public choices (train, bus, plane) fall into one group and the private choice (automobile) is in its own group. In other words, the public choices would tend to share unobserved characteristics that are not shared by the private choice.

Here's how to specify such a model in PROC MDC:

```
PROC MDC DATA=travel2;
  MODEL choice = ttme time cost / CHOICE=(mode)
    TYPE=NLOGIT;
  UTILITY U(1,)=ttme time cost;
  NEST LEVEL(1)=(1 2 3 @1, 4@2),
       LEVEL(2)=(1 2@1);
  ID id;
RUN;
```

Obviously, TYPE=NLOGIT requests the nested logit model. Such models require both a NEST statement and a UTILITY statement. When the predictor variables are the same for all choice options, the UTILITY statement takes the simple form shown here, with U(1,)= followed by a list of predictors in the model. The NEST statement must have at least two LEVELS. The LEVEL(1) option says that group 1 is composed of CHOICE values 1, 2, and 3, and group 2 is composed of CHOICE value 4. The LEVEL(2) option says that group 1 and group 2 are members of a single higher level group.

Results are shown in Output 7.14. As in the multivariate probit model, the *p*-value for COST is well above any conventional level of significance. TTME and TIME have *p*-values and coefficients that are similar to those in earlier models, although the coefficient for TTME is about twice as large as the one reported for the conditional logit model (Output 7.3).

Output 7.14 A Nested Logit Model with PROC MDC

Parameter Estimates					
Parameter	DF	Estimate	Standard Error	t Value	Approx Pr > |t|
TTME_L1	1	-0.0748	0.0101	-7.39	<.0001
TIME_L1	1	-0.003952	0.000719	-5.50	<.0001
COST_L1	1	-0.004363	0.005341	-0.82	0.4140
INC_L2G1C1	1	0.5601	0.1695	3.30	0.0010
INC_L2G1C2	1	1.3043	0.3507	3.72	0.0002

The two additional parameters allow for within-group correlations. Ideally, these parameters will be between 0 and 1, with 1 implying no correlation and 0 implying the highest correlation. When both parameters are equal to 1, the model reduces to the conditional logit model. The *t*-statistics for these parameters are not very helpful since they are tests of whether the parameters are equal to 0. What we really want are tests that the parameters are equal to 1. For INC_L2G1C1, which refers to the public transportation group, a 95 percent confidence interval does not include 1, which is evidence that there is a within-group correlation. The INC_L2G1C2 estimate is greater than 1, which is not consistent with the assumptions of the nested logit model. On the other hand, a 95 percent confidence interval easily includes 1, so the problematic estimate could certainly be attributable to sampling error.

Since the conditional logit model is a special case of the nested logit model, we can compare them by way of a likelihood ratio test. The log-likelihood for the nested model is -219.38 while the log-likelihood for the conditional model is -246.86. Twice the positive difference gives a chi-square of 54.96 with 2 d.f. (for the two additional parameters), which is highly significant. So we can reject the conditional model in favor of the nested model, which allows for departures from the IIA assumption. Also, the BIC (Schwartz) criterion for the nested model is 460, which is a better fit than the HEV model (BIC=483).

The nested logit model can get a lot more complicated, with multiple levels of nesting and predictors that describe different levels. A word of warning, however: Silberhorn et al. (2008) pointed out that there are actually two different versions of the nested logit model: the utility maximization nested logit and the non-normalized nested logit. PROC

MDC estimates the non-normalized nested model. Silberhorn et al. argue that the non-normalized model is inconsistent with random utility theory, and that the results it produces may sometimes be misleading.

Chapter 8

Logit Analysis of Longitudinal and Other Clustered Data

8.1 Introduction

In previous chapters, we assumed that all observations are independent—that is, the outcome for each observation is completely unrelated to the outcome for every other observation. While that assumption is quite appropriate for most data sets, there are many applications where the data can be grouped into natural or imposed clusters with observations in the same cluster tending to be more alike than observations in different clusters. Longitudinal data is, perhaps, the most common example of clustering. If we record an individual's responses at multiple points in time, we ordinarily expect those observations to be positively correlated. But there are many other applications in which the data have a cluster structure. For example, husbands and wives are clustered in families, and students are clustered in classrooms. In studies involving matching, each match group is a cluster. And, of course, cluster sampling naturally produces clustered observations.

There are several reasons why logistic regression should take clustering into account:

- Ignoring the clustering and treating the observations as though they are independent usually produces standard errors that are underestimated and test statistics that are overestimated. All methods for dealing with clustered data are designed to correct this problem.
- Conventional logistic regression applied to clustered data produces coefficient estimates that are not efficient. That means that there are, in principle, other methods whose coefficients have smaller true standard errors.
- In addition to these problems, clustered data also presents opportunities to correct for biases that may occur with any application of binary regression. One method for clustered data can correct for heterogeneity shrinkage. Another method can correct for bias due to omitted explanatory variables.

In this chapter, I'll discuss five different methods for clustered data that can be implemented in SAS:

- Robust standard errors with PROC GENMOD and PROC SURVEYLOGISTIC
- Generalized estimating equations (GEE) with PROC GENMOD
- Conditional logit (fixed-effects) models with PROC LOGISTIC
- Mixed models (random effects) with PROC GLIMMIX
- A hybrid method that combines fixed-effects with mixed models

8.2 Longitudinal Example

Let's begin with an example of longitudinal data, also known as panel data. The sample consisted of 316 people who survived residential fires in the Philadelphia area (Keane et al. 1996). They were interviewed at 3 months, 6 months, and 12 months after the fire. For each interview, the outcome variable PTSD is coded 1 if the person had symptoms of post-traumatic stress disorder, and 0 otherwise. The explanatory variables measured at each interview are

CONTROL	A scale of a person's perceived control over several areas of life.
PROBLEMS	The total number of problems reported in several areas of life.
SEVENT	The number of stressful events reported since the last interview.

In addition, there is one variable that was measured only at the initial interview:

COHES	A scale of family cohesion.

These are only a small fraction of the variables in the original study, and they were chosen solely for the sake of example. We shall ignore difficult questions of causal ordering here and simply assume that the dependent variable PTSD does not have any effect on the other four variables. For the moment, we'll also assume that any effects are contemporaneous so we don't have to worry about lags and leads.

Let's formulate a simple model. Let p_{it} be the probability that person *i* had PTSD symptoms at time *t*. Let $\mathbf{x}_{it}$ be a vector of explanatory variables for person *i* at time *t*. This vector can contain variables that differ at each interview and variables that are constant over interviews. It may also contain dummy variables to differentiate the three different interview times. Our initial model is just an ordinary logistic regression model with subscripts added for time.

$$\log\left(\frac{p_{it}}{1-p_{it}}\right)=\boldsymbol{\beta}\mathbf{x}_{it}\ . \qquad (8.1)$$

We'll first estimate this model *without* any special treatment for dependence among the observations. To accomplish this, the data is structured with a separate record for each person at each of the three time points, for a total of 948 records. Each record contains the variables described above, with COHES duplicated across the three records for each person. There is also a variable TIME equal to 1, 2, or 3 depending on whether the interview occurred at 3, 6, or 12 months after the fire. Finally, there is a variable called SUBJID, which contains a unique identification number for each *person*. We won't use that variable now, but it will be essential for the methods introduced later. The first 15 cases from the data set are shown in Output 8.1.

Output 8.1 First 15 Cases in PTSD Data Set

Obs	subjid	ptsd	control	problems	sevent	cohes	time
1	15	0	3.22222	5.625	1	8	1
2	15	0	3.16667	5.375	0	8	2
3	15	0	3.27778	3.750	1	8	3
4	18	1	2.55556	9.250	0	8	1
5	18	0	3.44444	4.375	0	8	2
6	18	0	3.33333	2.375	0	8	3
7	19	1	2.72222	7.750	1	7	1
8	19	1	2.77778	7.750	1	7	2
9	19	0	2.77778	7.500	1	7	3
10	23	0	3.38889	7.750	0	8	1
11	23	1	3.27778	7.250	1	8	2
12	23	0	3.33333	6.250	0	8	3
13	24	0	3.16667	6.875	0	9	1
14	24	0	3.88889	6.875	0	9	2
15	24	0	3.94444	4.375	0	9	3

A LOGISTIC program for estimating the model is:

```
PROC LOGISTIC DATA=ptsd;
  CLASS time / PARAM=GLM;
  MODEL ptsd(EVENT='1') = control problems sevent cohes time;
RUN;
```

Results in Output 8.2 show highly significant effects of four explanatory variables. The coefficients labeled TIME 1 and TIME 2 suggest that the odds of PTSD decline with time, but the differences are not statistically significant.

Output 8.2 PROC LOGISTIC Output for PTSD Data

Analysis of Maximum Likelihood Estimates						
Parameter		DF	Estimate	Standard Error	Wald Chi-Square	Pr > ChiSq
Intercept		1	1.4246	0.8287	2.9554	0.0856
control		1	-0.9594	0.2047	21.9760	<.0001
problems		1	0.2956	0.0505	34.3081	<.0001
sevent		1	0.3557	0.0804	19.5912	<.0001
cohes		1	-0.1782	0.0373	22.8583	<.0001
time	1	1	0.3566	0.2055	3.0116	0.0827
time	2	1	0.2499	0.2041	1.4993	0.2208
time	3	0	0	.	.	.

The problem with this analysis is that it assumes that the three observations for each person are independent. That's unlikely at face value and contradicted by the data. As Table 8.1 shows, the association between PTSD at time 1 and time 2 is quite strong, and even stronger between time 2 and time 3.

Table 8.1 Association between Symptoms of PTSD at Adjacent Time Points

PTSD at Time 1	%With PTSD at Time 2	PTSD at Time 2	%With PTSD at Time 3
Yes	47%	Yes	51%
No	18%	No	10%

Of course, the real question is whether this association persists after controlling for the explanatory variables in the model. To answer that question, I included PTSD at time 1 in a logistic model predicting PTSD at time 2, along with the other explanatory variables that are in Output 8.2. The coefficient for PTSD at time 1 was highly significant, indicating dependence over time (output not shown).

What are the consequences? Even with dependence among the observations, the coefficients in Output 8.2 should be consistent estimates of the true coefficients, and, therefore, approximately unbiased. They are not efficient, however. That means that there are other ways of estimating the coefficients that make better use of the data, thereby producing estimates with less sampling variability. A more serious problem is that the standard errors are likely to be underestimates of the true standard errors. And because the formula for the Wald chi-squares has the standard error in the denominator, the chi-squares likely to be overestimates. So, maybe the *p*-values aren't as small as they appear.

8.3 Robust Standard Errors

For the moment, let's ignore the lack of efficiency in the coefficient estimates and focus on the more serious problem of biased estimates of the standard errors. SAS has two procedures, SURVEYLOGISTIC and GENMOD, that can estimate a logistic regression model with "robust" standard errors that correct for dependence among the repeated measurements. These two procedures use a method that is variously known as the Huber-White method, the sandwich method, or the Taylor series expansion method. For some SAS procedures (e.g., GENMOD and GLIMMIX), robust standard errors are described in program code and output as *empirical standard errors*. Without going into the technical details, the basic idea is to calculate correlations among the Pearson residuals for the response variable within individuals, and then use those correlations to appropriately adjust the standard errors.

I'll first consider PROC SURVEYLOGISTIC, which is designed to handle complex sample surveys with stratification, clustering, and weighting. Repeated measures can be regarded as just a form of clustering (of the repeated measures within each individual), so we'll use the CLUSTER statement to request the robust standard errors.

Both the syntax and output of SURVEYLOGISTIC are quite similar to what we have previously seen for PROC LOGISTIC. Here is the code for estimating the logistic model for the PTSD data:

```
PROC SURVEYLOGISTIC DATA=ptsd;
  CLASS time / PARAM=GLM;
  MODEL ptsd(EVENT='1') = control problems sevent cohes time;
  CLUSTER subjid;
RUN;
```

On the CLUSTER statement, we simply name the variable that contains the ID numbers for the persons (which are the clusters in this case). If the CLUSTER statement were omitted, we would get the same results that we just saw in Output 8.2. Results *with* the CLUSTER statement are shown in Output 8.3.

Output 8.3 PROC SURVEYLOGISTIC Output with Robust Standard Errors

Analysis of Maximum Likelihood Estimates						
Parameter		**DF**	**Estimate**	**Standard Error**	**Wald Chi-Square**	**Pr > ChiSq**
Intercept		1	1.4246	0.9065	2.4695	0.1161
control		1	-0.9594	0.2281	17.6846	<.0001
problems		1	0.2956	0.0517	32.6886	<.0001
sevent		1	0.3557	0.0904	15.4726	<.0001
cohes		1	-0.1782	0.0469	14.4646	0.0001
time	**1**	1	0.3566	0.1847	3.7278	0.0535
time	**2**	1	0.2499	0.1728	2.0906	0.1482
time	**3**	0	0	.	.	.

Comparing Output 8.2 with Output 8.3, we see that the coefficients are identical. What *has* changed are the standard errors, which are somewhat larger in Output 8.3 for the first four variables. And since the chi-squares are calculated by dividing the parameter estimates by their standard errors, the chi-squares are somewhat lower than they were before. This doesn't make a noticeable difference in the *p*-values, however, because they were so low to begin with.

The exceptions to this pattern are the robust standard errors for the two indicator variables for TIME, which are smaller than the conventional standard errors. But that's consistent with a more general pattern: with robust standard errors, you tend to see the biggest percentage increase for variables that don't change over time (like COHES) and somewhat smaller increases for variables that do change over time (like CONTROL, PROBLEMS, and SEVENT). For variables that change over time but do not vary across persons (like TIME itself), the robust standard errors are often smaller than conventional standard errors.

Now let's see how to get the same results with PROC GENMOD, which is designed to estimate generalized linear models. These include linear models, logistic models, and the models for count data that we will examine in the next chapter. Here's the GENMOD code for the PTSD data:

```
PROC GENMOD DATA=ptsd DESC;
  CLASS subjid time;
  MODEL ptsd = control problems sevent cohes time /
    DIST=BINOMIAL TYPE3;
  REPEATED SUBJECT=subjid;
RUN;
```

Like PROC LOGISTIC, GENMOD models the *lower* value of the dependent variable by default. Unlike LOGISTIC, GENMOD does not accept the options EVENT='1' or DESCENDING on the MODEL statement. Instead, the DESCENDING option (here abbreviated DESC) must be placed on the PROC statement. Because GENMOD can estimate models for many different kinds of response variables, we need to specify the distribution of PTSD by using the option DIST=BINOMIAL, which can be abbreviated D=B. As with PROC LOGISTIC, the default link function for the binomial distribution is the logit link, but that can be overridden by choosing LINK=PROBIT or LINK=CLOGLOG. I've also included the very useful TYPE3 option because it gives overall tests for CLASS variables, tests that don't depend on the parameterization of those variables.

The REPEATED statement plays the same role as the CLUSTER statement in PROC SURVEYLOGISTIC. Without it, we would get maximum likelihood estimates with conventional standard errors. Unlike SURVEYLOGISTIC, however, the SUBJID variable must be specified not only on the REPEATED statement but also on the CLASS statement. If you forget and leave it off the CLASS statement (which I've done many times), you'll get an error message.

Output 8.4 PROC GENMOD Output with Robust Standard Errors

Analysis Of GEE Parameter Estimates							
Empirical Standard Error Estimates							
Parameter		**Estimate**	**Standard Error**	**95% Confidence Limits**		**Z**	**Pr > |Z|**
Intercept		1.4246	0.9022	-0.3438	3.1929	1.58	0.1143
control		-0.9594	0.2270	-1.4044	-0.5144	-4.23	<.0001
problems		0.2956	0.0515	0.1947	0.3964	5.74	<.0001
sevent		0.3557	0.0900	0.1793	0.5321	3.95	<.0001
cohes		-0.1782	0.0466	-0.2696	-0.0868	-3.82	0.0001
time	1	0.3566	0.1838	-0.0037	0.7169	1.94	0.0524

Analysis Of GEE Parameter Estimates							
Empirical Standard Error Estimates							
Parameter		Estimate	Standard Error	95% Confidence Limits		Z	Pr > |Z|
time	2	0.2499	0.1720	-0.0872	0.5870	1.45	0.1463
time	3	0.0000	0.0000	0.0000	0.0000	.	.

Score Statistics For Type 3 GEE Analysis			
Source	DF	Chi-Square	Pr > ChiSq
control	1	16.39	<.0001
problems	1	29.63	<.0001
sevent	1	11.84	0.0006
cohes	1	14.32	0.0002
time	2	4.05	0.1319

Results are shown in Output 8.4. Although the heading says “Analysis of GEE Parameter Estimates,” the estimates reported here are actually just conventional maximum likelihood estimates. In the next section, we’ll see how to get GEE estimates that correct for dependence. The subheading says “Empirical Standard Error Estimates,” which is GENMOD’s way of saying robust standard errors. The standard errors are a little different from the ones reported by PROC SURVEYLOGISTIC because of slight differences in how the formulas adjust for the number of parameters relative to the number of cases. Note that GENMOD reports z-statistics rather than chi-square statistics. But if you square them, you get chi-square statistics, and the p-values are the same either way.

One downside of PROC GENMOD is that it doesn’t report odds ratios. You can request them with an ESTIMATE statement, but you’ll need a different statement for each odds ratio. For example, to get the odds ratio for SEVENT in Output 8.4, include the following statement after the MODEL statement:

```
ESTIMATE 'sevent' sevent 1 / EXP;
```

The ESTIMATE statement requires a label, which can be any text enclosed by quotes. This statement says to take the coefficient of SEVENT, multiply it by 1, and then exponentiate it. Results are shown in Output 8.5. The odds ratio of 1.4272 is shown on the second line of the column “L'Beta Estimate”. We can interpret this by saying that “each additional stressful event increases the odds of PTSD by 43 percent." The last line of the output gives the 95 percent confidence interval for the odds ratio. You can have as many ESTIMATE statements as you wish.

Output 8.5 *Odds Ratio Produced by Estimate Statement*

Contrast Estimate Results

Label	Mean Estimate	Mean Confidence Limits		L'Beta Estimate	Standard Error	Alpha	L'Beta Confidence Limits		Chi-Square	Pr > ChiSq
sevent	0.5880	0.5447	0.6300	0.3557	0.0900	0.05	0.1793	0.5321	15.62	<.0001
Exp(sevent)				1.4272	0.1285	0.05	1.1964	1.7025		

8.4 GEE Estimation with PROC GENMOD

Robust standard errors solve the biggest problem that arises from repeated observations on the same individual. But the coefficient estimates are still not efficient, which implies that their *true* standard errors are higher than they need to be. In principle, one ought to be able to get better estimates, in the sense of having less sampling variability. One way to accomplish this is the method of generalized estimating equations (Diggle et al. 1994).

Here's a brief, nontechnical explanation of this method. In ordinary logistic regression, the maximum likelihood estimates can be obtained by an algorithm known as iteratively reweighted least squares. What that means is that each step in the algorithm is accomplished by weighted least squares, with both the weights and the constructed dependent variable changing at each iteration as functions of the results at the last iteration. In the matrix formulation of weighted least squares, there is an $N \times N$ weight matrix **W**, which has off-diagonal elements equal to 0 and diagonal elements equal to $p_i(1-p_i)$, where p_i is the predicted probability from the previous iteration. The GEE algorithm extends this approach to do iterative *generalized* least squares. In this method, the matrix **W** has nonzero off-diagonal elements that are functions of the correlations among the observations. These correlations are re-estimated at each iteration based on correlations among the Pearson residuals.

With GENMOD, GEE estimation is invoked by specifying a correlation structure on the REPEATED statement. Here is the program for our PTSD example:

```
PROC GENMOD DATA=ptsd DESC;
  CLASS subjid time;
  MODEL ptsd = control problems sevent cohes time / D=B;
  REPEATED SUBJECT=subjid / WITHIN=time TYPE=UN CORRW;
RUN;
```

As before, the SUBJECT option on the REPEATED statement names a variable that contains a unique identification code for each cluster. It is assumed that observations are independent between clusters and correlated within clusters. The WITHIN option names a variable that distinguishes different items within a cluster. In this example, the items are distinguished by different times. This variable must also be named in the CLASS statement. The WITHIN option can be omitted if either (a) each cluster has the same number of cases and the data is appropriately sorted within the cluster (as in this example) or (b) the items in each cluster are all treated as equivalent (as with students in a classroom).

The TYPE option is used to specify the structure of the correlation matrix among the observations within each cluster. In this case, I've chosen UN for *unstructured*, which means that no restrictions are placed on the correlations. This is probably the best choice if the number of time points is small. With three time points, there are three correlations to be estimated: 1 with 2, 2 with 3, and 1 with 3. As the number of time points gets larger, the number of correlations increases rapidly. So with a large number of time points, you may want to impose some structure to avoid having to estimate so many correlations (which can reduce the power of statistical tests and often results in non-convergence or out-of-memory conditions). We'll see how to do this shortly. The CORRW option simply asks PROC GENMOD to print out the estimate of the "working" correlation matrix.

Results are shown in Output 8.6. We first see estimates of the over-time correlations, which are produced by the CORRW option. The estimated correlation between PTSD at time 1 and PTSD at time 2 is .1891. Between PTSD at time 1 and PTSD at time 3, it's .2583. And .3878 is the correlation for time 2 and time 3. Actually, these are residual correlations that partial out the effects of the explanatory variables. If we fit a model with no explanatory variables, the correlations are a little higher (not shown). Somewhat surprisingly, the time 1 response is more highly correlated with time 3 than with time 2.

Output 8.6 GEE Results for PTSD Data with an Unstructured Correlation Matrix

Working Correlation Matrix			
	Col1	Col2	Col3
Row1	1.0000	0.1891	0.2538
Row2	0.1891	1.0000	0.3878
Row3	0.2538	0.3878	1.0000

GEE Fit Criteria	
QIC	984.7186
QICu	983.0397

Analysis Of GEE Parameter Estimates Empirical Standard Error Estimates							
Parameter		Estimate	Standard Error	95% Confidence Limits		Z	Pr > |Z|
Intercept		1.6078	0.8689	-0.0952	3.3108	1.85	0.0643
control		-0.9071	0.2159	-1.3302	-0.4840	-4.20	<.0001
problems		0.2559	0.0501	0.1577	0.3540	5.11	<.0001
sevent		0.2740	0.0867	0.1041	0.4439	3.16	0.0016
cohes		-0.1911	0.0455	-0.2803	-0.1018	-4.20	<.0001
time	1	0.4164	0.1781	0.0673	0.7656	2.34	0.0194
time	2	0.2717	0.1664	-0.0544	0.5978	1.63	0.1024
time	3	0.0000	0.0000	0.0000	0.0000	.	.

The next section of Output 8.6 displays the "GEE Fit Criteria." I'll hold off discussing those statistics until we consider some alternative models. In the last panel, we see the coefficients and their associated statistics. As expected, the coefficients are all different from the conventional maximum likelihood estimates in Output 8.4, but not markedly different. This is usually the case, although sometimes you will see more dramatic differences. Now let's try a structured correlation matrix. One of the simplest models is the "exchangeable" structure, which is specified as TYPE=EXCH on the REPEATED statement. This produced the results shown in Output 8.7. As you can see from the working correlation matrix, this option specifies that the correlations are equal for any pair of time points, a fairly strong assumption for longitudinal data. (I used the CORRW option so that the structure of the correlations would be clear. But you actually don't need it because the single correlation is automatically reported with the EXCH option.) Both the coefficients and the standard errors change slightly under the exchangeable specification.

Output 8.7 GEE Results for an Exchangeable Correlation Structure

Working Correlation Matrix			
	Col1	Col2	Col3
Row1	1.0000	0.2727	0.2727
Row2	0.2727	1.0000	0.2727
Row3	0.2727	0.2727	1.0000

Exchangeable Working Correlation	
Correlation	0.2727435633

GEE Fit Criteria	
QIC	984.3825
QICu	982.8269

Analysis Of GEE Parameter Estimates Empirical Standard Error Estimates							
Parameter		Estimate	Standard Error	95% Confidence Limits		Z	Pr > |Z|
Intercept		1.7927	0.8619	0.1033	3.4820	2.08	0.0375
control		-0.9601	0.2147	-1.3809	-0.5393	-4.47	<.0001
problems		0.2497	0.0497	0.1523	0.3471	5.02	<.0001
sevent		0.2810	0.0864	0.1116	0.4503	3.25	0.0011
cohes		-0.1871	0.0451	-0.2755	-0.0987	-4.15	<.0001
time	1	0.4100	0.1782	0.0607	0.7593	2.30	0.0214
time	2	0.2699	0.1662	-0.0558	0.5955	1.62	0.1043
time	3	0.0000	0.0000	0.0000	0.0000	.	.

The standard errors in Output 8.6 and Output 8.7 are labeled "Empirical Standard Error Estimates," which means that they are robust estimates that do not depend on the correctness of the structure imposed on the working correlation matrix. It's also possible to get *model-based* estimates of standard errors by putting the option MODELSE on the REPEATED statement. These standard errors, which are based directly on the assumed correlation structure, are then used in constructing confidence intervals and the *z*-statistics. The model-based standard errors ought to be better estimates if the assumed model for the correlation structure is correct, but worse if the assumed model is incorrect. It's certainly safer to stick with the robust estimates. On the other hand, the robust standard errors may have less power in small samples, especially as the number of time points gets larger.

The exchangeable structure is often too restrictive, because it doesn't allow for a pattern that is very common in longitudinal data: measurements that are close together in time tend to have higher correlations than measurements that are farther apart. This sort of pattern can be modeled by specifying TYPE=AR, which invokes a first-order autoregressive structure. But this typically makes the correlations decline *too* rapidly with distance; specifically, if r is the correlation between measurements that are one time unit apart, then the correlation between measures that are two time units apart must be r^2, the correlation between measurements that are three time units apart must be r^3, and so on.

My favorite correlation structure for longitudinal data is TYPE=MDEP(n), where n can be any number less than the number of time points. I recommend choosing n to be one less than the number of time points. So for our example, that would be TYPE=MDEP(2). Output 8.8 shows the results from fitting the MDEP model. This correlation structure says that there is one correlation for measurements that are one time unit apart (here .2849), another correlation for measurements that are two time units apart (here .2547), and another correlation for measurements are three time units apart (which doesn't apply here).

Output 8.8 GEE Results for an MDEP(2) Correlation Structure

Working Correlation Matrix			
	Col1	Col2	Col3
Row1	1.0000	0.2849	0.2547
Row2	0.2849	1.0000	0.2849
Row3	0.2547	0.2849	1.0000

GEE Fit Criteria	
QIC	984.4836
QICu	982.9455

Analysis Of GEE Parameter Estimates							
Empirical Standard Error Estimates							
Parameter		Estimate	Standard Error	95% Confidence Limits		Z	Pr > |Z|
Intercept		1.7744	0.8608	0.0873	3.4615	2.06	0.0393
control		-0.9534	0.2147	-1.3741	-0.5326	-4.44	<.0001
problems		0.2493	0.0495	0.1522	0.3463	5.03	<.0001
sevent		0.2780	0.0863	0.1089	0.4472	3.22	0.0013
cohes		-0.1871	0.0451	-0.2755	-0.0987	-4.15	<.0001
time	1	0.4121	0.1783	0.0628	0.7615	2.31	0.0208
time	2	0.2705	0.1660	-0.0549	0.5958	1.63	0.1032
time	3	0.0000	0.0000	0.0000	0.0000	.	.

How do you decide which is the right correlation structure? Well, as I said before, if the number of time points is relatively small (say, four or less), I would definitely choose the unstructured matrix. If the number of time points is larger, I would first try to fit the unstructured matrix. If it works (and sometimes it won't), I would examine the correlations to see if there's a trend toward correlations that decline with the time between measurements. If there is such a trend, I'd be inclined to go with the MDEP structure. If not, the exchangeable structure might be fine. Incidentally, if you're using GEE to handle other kinds of clustering, like students within schools or people within neighborhoods, the exchangeable structure is usually the only way to go because there's no natural ordering of the people within clusters. The first-order autoregressive (AR) structure is rarely appropriate unless the correlations are all quite high.

One can also use the QIC statistic, shown in the three preceding outputs, to compare the empirical fit for different correlation structures. This statistic was introduced by Pan (2001) as the GEE equivalent of the AIC statistic, which is often used to compare models estimated by maximum likelihood. As with the AIC statistic, models with lower values of QIC are preferred. The other fit statistic (QICu) is an approximation to the QIC that is recommended only for comparing models with different sets of predictors, not for comparing correlation structures. For the three correlation structures that we have examined here, the QIC statistics are all about the same. So it's not very helpful for this example. In general, my experience with this statistic has not been encouraging. It sometimes indicates a preference for structures that are clearly inconsistent with the unstructured correlation matrix. Hin and

Wang (2009) have proposed an alternative statistic that seems to perform much better, at least according to their simulations. But that statistic (which they call CIC) is not currently available in SAS.

One potential criticism of the GEE models we have seen so far is that they are based on Pearson correlations for the residuals of a dichotomous response. Those correlations should vary, in part, as a function of the predicted value of the dichotomous variable. Predicted values near 0 or 1 should lead to lower correlations. It might be preferable to specify the model in terms of measures of association that do not depend on the predicted values.

GENMOD offers a solution to that problem by parameterizing the over-time associations in terms of odds ratios rather than correlations. The method is called “alternating logistic regression” because the iterative algorithm alternates between (1) a GEE step for the main model and (2) a conventional logistic regression for the odds ratios that describe the relationships among the repeated measures. GENMOD allows for some fairly complex models for these odds ratios, such as different values for different subgroups, and odds ratios that depend on other explanatory variables. Here, we’ll look at two of the simpler models that are analogs of the TYPE=UN and TYPE=EXCH.

To do the odds-ratio equivalent of the unstructured model for the PTSD data, we change the REPEATED statement to read

```
REPEATED SUBJECT=subjid / WITHIN=time LOGOR=FULLCLUST;
```

LOGOR refers to the logarithm of the odds ratio, and FULLCLUST says that there is a full set of odds ratios within each cluster. So this option allows for a different (partial) odds ratio for each pair of repeated outcome measurements. Results are shown in Output 8.9.

Output 8.9 GEE Results for an Unstructured Odds Ratio Model

Log Odds Ratio Parameter Information	
Parameter	**Group**
Alpha1	(1, 2)
Alpha2	(1, 3)
Alpha3	(2, 3)

Analysis Of GEE Parameter Estimates							
Empirical Standard Error Estimates							
Parameter		Estimate	Standard Error	95% Confidence Limits		Z	Pr > |Z|
Intercept		1.5381	0.8748	-0.1765	3.2527	1.76	0.0787
control		-0.9365	0.2167	-1.3612	-0.5118	-4.32	<.0001
problems		0.2735	0.0502	0.1752	0.3718	5.45	<.0001
sevent		0.2885	0.0865	0.1189	0.4580	3.33	0.0009
cohes		-0.1843	0.0462	-0.2750	-0.0937	-3.99	<.0001
time	1	0.4099	0.1817	0.0537	0.7661	2.26	0.0241
time	2	0.2799	0.1712	-0.0557	0.6156	1.63	0.1021
time	3	0.0000	0.0000	0.0000	0.0000	.	.
Alpha1		0.7762	0.2966	0.1949	1.3576	2.62	0.0089
Alpha2		1.2778	0.3382	0.6149	1.9407	3.78	0.0002
Alpha3		1.6549	0.3632	0.9431	2.3668	4.56	<.0001

The "Log Odds Ratio Parameter Information " table defines the parameters ALPHA1 through ALPHA3 corresponding to the pairs of repeated measurements. The estimates of these parameters then appear as part of the main regression table in Output 8.9. Keep in mind that these are logarithms of the odds ratios, not the odds ratios themselves. What's nice is that we get significance tests for each of these estimates, something that we couldn't get with conventional GEE. The coefficients for the explanatory variables are very similar to what we saw in Output 8.7.

To estimate the analog of the exchangeable model, we change the REPEATED statement to read

```
REPEATED SUBJECT=subjid / WITHIN=time LOGOR=EXCH;
```

which says that there is only a single odds ratio for all pairs of repeated measurements. That produces the results in Output 8.10. The estimates and associated statistics are not much different from those in the preceding output, but now we only get a single log-odds ratio of 1.206. Exponentiating this value gives an odds ratio of 3.34. Thus, for example, if you had PTSD at time 1, your odds of having PTSD at time 3 are 3.34 what they would be if you did not have PTSD at time 1 (controlling for other variables in the model).

Output 8.10 GEE Results for an Exchangeable Odds Ratio Model

Analysis Of GEE Parameter Estimates Empirical Standard Error Estimates							
Parameter		Estimate	Standard Error	95% Confidence Limits		Z	Pr > |Z|
Intercept		1.7035	0.8703	-0.0022	3.4093	1.96	0.0503
control		-0.9728	0.2165	-1.3972	-0.5484	-4.49	<.0001
problems		0.2642	0.0495	0.1672	0.3613	5.33	<.0001
sevent		0.2922	0.0868	0.1221	0.4622	3.37	0.0008
cohes		-0.1825	0.0460	-0.2727	-0.0923	-3.97	<.0001
time	1	0.4079	0.1810	0.0532	0.7626	2.25	0.0242
time	2	0.2679	0.1703	-0.0659	0.6017	1.57	0.1157
time	3	0.0000	0.0000	0.0000	0.0000	.	.
Alpha1		1.2060	0.2314	0.7524	1.6596	5.21	<.0001

8.5 Mixed Models with PROC GLIMMIX

As we have seen, the principal attraction of GEE is the fact that it produces efficient coefficient estimates that have minimum sampling variability. Mixed modeling has the same attraction, but also offers a number of additional features. In this approach, clustering is treated as a random effect in a mixed model—so named because it includes both fixed and random effects. Although mixed models are quite similar to GEE, they have two potential advantages. First, much more complex models are possible, with multiple levels of clustering, overlapping clusters, and random coefficients. Second, estimation of mixed models can correct for heterogeneity shrinkage discussed in Section 3.12. In other words, mixed models are subject specific rather than population averaged like the models estimated by GEE (McCulloch 1997, Rodriguez and Goldman 1995).

Mixed models differ from GEE models in that dependence between repeated measurements is directly built into the model. In its simplest form, the mixed logit model can be written as follows:

$$\log\left(\frac{p_{it}}{1-p_{it}}\right)=\alpha_i+\boldsymbol{\beta}\mathbf{x}_{it}\,. \qquad (8.2)$$

This equation differs from equation (8.1) by the presence of α_i, which can be thought of as representing the combined effects of all unobserved variables that affect the outcome but which are stable over time. This is sometimes called the *random intercepts* model. And α_i is often said to represent unobserved heterogeneity. We shall assume that each α_i is a random variable with a specified probability distribution. Specifically, the α_i's are assumed to be independent of the $\mathbf{x}_{it}$ and to have a normal distribution with a mean of 0. In the simpler models, they are also assumed to be independent of each other and to have a constant variance of σ^2.

The presence of α_i in equation (8.2) *induces* a correlation between the repeated measurements of the outcome variable. If α_i is high for particular individual, then many of that individual's responses will tend to be 1. On the other hand, if α_i is low, then many of the individual's responses will tend to be 0. The greater the variance of α_i, the higher the within-individual correlations in the responses. This model implies a form of exchangeability: the association between repeated measurements at any two time points is the same for any pair of time points. Thus, it does not allow the association to diminish with the length of time between measurements.

SAS has two procedures that can estimate mixed logistic regression models, PROC NLMIXED and PROC GLIMMIX. Previously, I strongly preferred NLMIXED because it did true maximum likelihood estimation, while early versions of GLIMMIX only did something called pseudo-maximum likelihood. Pseudo ML is known to be inaccurate for dichotomous outcomes, especially when the number of observations per cluster is small (McCulloch 1997, Rodriguez and Goldman 1995). But now GLIMMIX can do true maximum likelihood, using the same methods for integrating the likelihood function that are found in PROC NLMIXED. We'll focus on GLIMMIX here because it has a much simpler syntax than NLMIXED. Here's a program for estimating the random intercepts model for the PTSD data:

```
PROC GLIMMIX DATA=ptsd METHOD=QUAD;
  CLASS subjid time;
  MODEL ptsd = control problems sevent cohes time / D=B
     SOLUTION;
  RANDOM INTERCEPT / SUBJECT=subjid;
  COVTEST 0;
RUN;
```

The syntax for GLIMMIX combines many of the features of PROC GENMOD and PROC MIXED (which does linear mixed models). The option METHOD=QUAD is essential for getting maximum likelihood estimates rather than pseudo likelihood estimates. QUAD refers to Gaussian quadrature, which is the method used to integrate the likelihood function over the distribution of the random intercept. The D=B option on the MODEL statement specifies a binomial distribution for the dependent variable, with a default logit link function. Note that, unlike LOGISTIC, SURVEYLOGISTIC, or GENMOD, PROC GLIMMIX does not require the DESCENDING option or the EVENT='1' option. With D=B, GLIMMIX models the probability of a 1 rather than a 0. The SOLUTION option (which can be abbreviated as just S) requests that coefficient estimates be reported, in addition to the usual hypothesis tests.

The RANDOM statement (which is much like the RANDOM statement in PROC MIXED), declares that the intercept will be treated as a random variable. The SUBJECT option says that this random variable will be different for each value of the SUBJID variable, but will be the same for observations with the same value of SUBJID. The COVTEST statement provides a test of the null hypothesis that the variance of α is 0. Output 8.11 gives the results.

Note that I've included SUBJID in the CLASS statement. Unlike GENMOD, GLIMMIX does not *require* that the subject variable be in the CLASS statement, but it's usually a good idea to do so. If the data are sorted by the subject variable, you'll get the same results whether that variable is or is not in the CLASS statement. But if the data are not sorted by the subject variable, leaving it out of the CLASS statement will yield completely erroneous results.

Output 8.11 PROC GLIMMIX Results for a Random Intercepts Model

Covariance Parameter Estimates			
Cov Parm	**Subject**	**Estimate**	**Standard Error**
Intercept	subjid	2.1242	0.5774

Solutions for Fixed Effects						
Effect	**time**	**Estimate**	**Standard Error**	**DF**	**t Value**	**Pr > |t|**
Intercept		2.3588	1.1838	314	1.99	0.0472
control		-1.3239	0.2969	627	-4.46	<.0001
problems		0.3545	0.07049	627	5.03	<.0001
sevent		0.4016	0.1086	627	3.70	0.0002
cohes		-0.2476	0.06249	627	-3.96	<.0001
time	1	0.5258	0.2518	627	2.09	0.0371
time	2	0.3513	0.2394	627	1.47	0.1428
time	3	0	.	.	.	.

Tests of Covariance Parameters Based on the Likelihood					
Label	**DF**	**-2 Log Like**	**ChiSq**	**Pr > ChiSq**	**Note**
Parameter list	1	966.85	41.54	<.0001	MI

MI: P-value based on a mixture of chi-squares.

The first thing we see in the output is an estimate of the variance of the random intercept: 2.1242. If this were 0, we'd be back to an ordinary logistic regression model. At the bottom of the output, we get a likelihood ratio chi-square test of the null hypothesis that this variance is 0. Clearly, that hypothesis can be rejected. Incidentally, there are serious technical difficulties that arise in the calculation of the *p*-value for a likelihood ratio test that a variance is equal to 0. But GLIMMIX solves those problems by calculating *p*-values "based on a mixture of chi-squares."

In the "Solutions for Fixed Effects" table, we find the coefficients, standard errors, *t*-statistics and *p*-values. Let's compare these results in with those in Output 8.10 for the GEE exchangeable odds ratio model, which is the GEE model that's most similar to the random intercepts model. The *t*-statistics in Output 8.11 are all pretty close to the *z*-statistics in Output 8.10 (both are just ratios of the coefficients to their standard errors). But the coefficients for the random intercepts model are all larger in magnitude than the coefficients for the GEE model. That's to be expected because mixed model coefficients correct for heterogeneity shrinkage but GEE coefficients do not. The greater the variance of α, the bigger the difference you will find between mixed model and GEE coefficients.

So if you really want subject-specific estimates of the coefficients, mixed models are the way to go. But mixed models can be much more computationally intensive than GEE because of the need for numerical integration. And not infrequently, they break down completely for reasons that are not always apparent. If your main goal is to get accurate *p*-values, GEE may do perfectly well.

Mixed models can be elaborated in several different ways, one of which is to allow the coefficients themselves to vary randomly across persons. This is sometimes called a random slopes model. Let's try it for the PTSD data. Suppose we think that SEVENT (the number of stressful events) has different effects for different people. To allow for this, we can modify the model by putting an *i* subscript on the coefficient vector.

$$\log\left(\frac{p_{it}}{1-p_{it}}\right)=\alpha_i+\boldsymbol{\beta}_i\mathbf{x}_{it}\,. \tag{8.3}$$

This equation would allow *all* the coefficients to vary across persons, but we want to focus on the special case where just one of them is allowed to vary. We'll assume that the SEVENT coefficient is normally distributed, with a mean and variance that we shall estimate. We'll also assume that the random coefficient is independent of everything else on the right-hand side of the equation. Here's the GLIMMIX code for implementing that model:

```
PROC GLIMMIX DATA=ptsd METHOD=QUAD;
  CLASS subjid time;
  MODEL ptsd = control problems sevent cohes time / D=B S;
  RANDOM INTERCEPT sevent/ SUBJECT=subjid;
  COVTEST 0 .;
  COVTEST . 0;
RUN;
```

To make a coefficient random, we merely put the variable name on the RANDOM statement. This can only be done for variables that vary within subject. We could not, for example, put COHES on the RANDOM statement because it was only measured once and does not change over time. Also, it's nearly always a bad idea to allow coefficients to be random without also specifying a random intercept.

Notice the two COVTEST statements. The first one tests the null hypothesis that the variance of the random intercept is 0, while allowing the random slope to be non-zero. The second tests the null hypothesis that the random slope is 0, while allowing the random intercept to be non-zero.

Results are shown in Output 8.12. In the first table, we see the estimated variances of the random intercept and the random slope. At first glance, the slope variance does not appear to be significantly different from 0 because it is not more than twice its standard error. Nevertheless, the more accurate likelihood ratio test reported in the bottom table yields a *p*-value of .015. So we do have evidence of variability in the effect of SEVENT across persons in the sample.

Output 8.12 PROC GLIMMIX Results for a Random Slope Model

Covariance Parameter Estimates			
Cov Parm	Subject	Estimate	Standard Error
Intercept	subjid	2.1546	0.6649
sevent	subjid	0.7009	0.4773

Solutions for Fixed Effects						
Effect	time	Estimate	Standard Error	DF	t Value	Pr > |t|
Intercept		2.8295	1.2997	314	2.18	0.0302
control		-1.4913	0.3313	387	-4.50	<.0001
problems		0.3774	0.07738	387	4.88	<.0001
sevent		0.3054	0.1521	240	2.01	0.0458
cohes		-0.2620	0.06785	387	-3.86	0.0001
time	1	0.6014	0.2720	387	2.21	0.0276
time	2	0.4236	0.2581	387	1.64	0.1016
time	3	0	.	.	.	.

Tests of Covariance Parameters Based on the Likelihood					
Label	DF	-2 Log Like	ChiSq	Pr > ChiSq	Note
Parameter list	1	948.35	27.76	<.0001	MI
Parameter list	1	925.31	4.71	0.0150	MI

MI: P-value based on a mixture of chi-squares.

The coefficients reported in the "Solutions for Fixed Effects" table are very similar to those in Output 8.11, which did not have the random slope. However, the coefficient for SEVENT (.3054) now has a different interpretation. Instead of being the unique coefficient of SEVENT, it is merely the mean of the coefficients across persons. With a standard deviation of .84 (the square root of the random-effects variance of .7009), there is a lot of variability in this effect. Under the assumption that the coefficient is normally distributed, we conclude that 68 percent of the people have coefficients that range between -.53 and 1.15 (adding and subtracting one standard deviation). That's a big range.

It's possible to have more than one random slope, but when I tried to introduce random slopes for each of the other variables, the algorithm failed to converge. The default is to assume that the all random effects are uncorrelated. But you can allow for correlations by specifying TYPE=UNR as an option on the RANDOM statement. When I did this for the model in Output 8.12, I got a correlation of -.29 between the random intercept and the random slope. But that correlation was not significantly different from 0.

By default, the standard errors reported by GLIMMIX are model-based rather than robust. This means that they are based on the exchangeability assumption, an assumption that is not easily relaxed in the mixed model framework. It's easy to switch to robust standard errors, however, by putting the option EMPIRICAL on the PROC statement. That's generally a good idea unless the sample size is small. (For smaller samples, the empirical standard errors can be somewhat improved by specifying EMPIRICAL=MBN.) For this example, switching to robust standard errors made very little difference.

8.6 Fixed-Effects with Conditional Logistic Regression

Although GEE and mixed models are wonderful tools, they don't correct for bias resulting from omitted explanatory variables at the cluster level. To be more specific, when you have multiple observations per person—as in the PTSD example—it is possible to statistically control for all stable characteristics of persons, regardless of whether those characteristics can be measured. Needless to say, this is an extremely attractive possibility, and one that is easily implemented with the LOGISTIC procedure. Unfortunately, as we shall see, there are significant limitations that keep this approach from being the universally preferred method.

The equation for the model is exactly the same as equation (8.2) for the random intercepts model:

$$\log\left(\frac{p_{it}}{1-p_{it}}\right)=\alpha_i+\boldsymbol{\beta}\mathbf{x}_{it}.$$

As before, the α_i term represents all unobserved differences among individuals that are stable over time. Because α_i is the same for a given person at all three time points, a positive correlation is induced among the observed outcomes.

In the mixed model approach, we treated α_i as a random variable with a specified probability distribution: normal with mean 0, variance σ^2, and independent of $\mathbf{x}_{it}$. In this section, we treat α_i as a set of fixed constants, one for each individual in the sample, an approach that is sometimes referred to as a fixed-effects model. But how can we estimate it? If the response variable were continuous, we could use ordinary least squares, with a dummy variable for each individual *i* (less one) (Allison 2005). For example, if the dependent variable were CONTROL instead of PTSD, we could fit a model to the 948 observations with 316 dummy variables. That may seem like an outrageous number of variables in the model, but it's perfectly legitimate. An equivalent method that is more computationally tractable is to transform all variables so that they are expressed as deviations from each individual's means. Then, the model is run on the transformed variables. PROC GLM can accomplish this automatically by using the ABSORB statement.

Unfortunately, the dummy variable or mean deviation methods aren't satisfactory for logistic regression, unless the number of clusters (dummy variables) is small and the number of observations per cluster is relatively large. It certainly wouldn't be appropriate for the PTSD example. The difficulty is caused by the "incidental parameters problem" (Kalbfleisch and Sprott 1970). In the asymptotic theory of maximum likelihood estimation, it is usually assumed that the number of observations gets large while the number of parameters to be estimated remains constant. However, when equation (8.2) is applied to longitudinal data, each additional individual adds an additional parameter to the model. This can lead to substantial overestimates of the coefficients for other variables.

One solution to the incidental parameters problem is called conditional likelihood estimation (Chamberlain 1980). In constructing the likelihood function, we condition on the number of 1's and 0's that are observed for each individual. For example, if a person had PTSD symptoms at time 1 but not at time 2 and time 3, we ask, "Given that this person had only one occasion with PTSD symptoms, what's the probability that it occurred at time 1 and not at time 2 or time 3?" We write an expression for this probability as a function of the explanatory variables and the β parameters, and we multiply these probabilities together for all individuals to get the overall likelihood. When this is done, the α_i parameters cancel from the likelihood function. We say that they have been *conditioned out* of the likelihood. Presto, no more incidental parameters problem.

How can we implement this method? The conditional likelihood function for the fixed-effects model has a form that's identical to the conditional logit model that we examined for discrete choice models in Chapter 7. And as we saw there, the conditional logit model can be implemented with the STRATA statement in PROC LOGISTIC. Here's the code for estimating the fixed-effects model for the PTSD data:

```
PROC LOGISTIC DATA=ptsd ;
  CLASS time / PARAM=GLM;
  MODEL ptsd(EVENT='1') = control problems sevent cohes time;
  STRATA subjid;
RUN;
```

Selected results are shown in Output 8.13.

Output 8.13 PROC LOGISTIC Output for Fixed-Effects Logit Model

Model Information	
Data Set	MY.PTSD
Response Variable	ptsd
Number of Response Levels	2
Number of Strata	316
Number of Uninformative Strata	181
Frequency Uninformative	543
Model	binary logit
Optimization Technique	Newton-Raphson ridge

Strata Summary				
Response Pattern	ptsd 0	ptsd 1	Number of Strata	Frequency
1	0	3	37	111
2	1	2	48	144
3	2	1	87	261
4	3	0	144	432

Analysis of Maximum Likelihood Estimates						
Parameter		DF	Estimate	Standard Error	Wald Chi-Square	Pr > ChiSq
control		1	-1.0983	0.4221	6.7700	0.0093
problems		1	0.2139	0.1027	4.3401	0.0372
sevent		1	0.2027	0.1343	2.2783	0.1312
cohes		0	0	.	.	.
time	1	1	0.7941	0.2972	7.1402	0.0075
time	2	1	0.4365	0.2597	2.8252	0.0928
time	3	0	0	.	.	.

In the "Model Information" table, we see that out of 316 strata (one for each person), 181 were deemed "uninformative." And since each of these 181 persons had three records, 543 of the observations were uninformative. To understand what this means, we must go to the "Strata Summary" table. There we find that 144 people did not have PTSD at any of the three time points, and another 37 people had PTSD at all three time points. These add up to the 181 uninformative strata, fully 57 percent of the sample. Why are they uninformative? Well, the conditional likelihood is trying to explain why a person with a certain number of occasions with PTSD symptoms had them at some times and not at other times. But if we know that a person had symptoms at all three interviews or at none of the interviews, there's no variation in timing to explain.

Several other things are evident in the LOGISTIC output: there's no intercept, there's no coefficient for COHES, and the *p*-values are a lot higher than those we have seen in previous regression output for these data. The reason we don't get a COHES coefficient is that COHES is constant over time. This is a quite general and often maddening characteristic of fixed-effects models: they can't produce coefficients for variables that don't vary within clusters. The explanation is both simple and intuitive. In constructing the conditional likelihood, we essentially pose the question, "Given that *k* events happened to this individual, what factors explain why they happened at these particular times and not at other times?" Clearly, variables that are constant over time cannot explain why events happened at some times and not others. That doesn't mean that the time-constant variables aren't controlled. As I noted earlier, the method actually controls for *all* time constant variables, not just those we happen to have measured. Or as it's sometimes stated, each individual serves as his or her own control.

The main reason the chi-squares are small is that the standard errors are substantially larger than they were for GEE and mixed model estimation. If you compare the coefficients, they're not all that different from those in, say, Output 8.10. So why are the standard errors larger? Because fixed-effects estimation discards a lot of data that's used by GEE or mixed models. As I just explained, any person whose PTSD values didn't change over the three periods is automatically excluded from the likelihood function. Even those who did have PTSD symptoms at some times and not at other times don't contribute as much information as they did in the GEE or mixed models. Only within-person variation is used, not between-person variation. The result is standard errors that are about twice as large as in the GEE or mixed models.

The contrast between fixed-effects and random-effects (mixed model) estimation is a classic trade-off between bias and inefficiency. Fixed-effects estimates are much less prone to bias because the conditional likelihood discards information that is potentially contaminated by confounding variables. On the other hand, if confounding variables are not a problem, then the fixed-effects approach can discard much potentially useful information. That seems to be the case in the PTSD example where the fixed-effects coefficients differ only slightly from the GEE or mixed model coefficients.

8.7 Postdoctoral Training Example

The techniques that work for panel data also work for other kinds of clustered data. In this section, we'll apply the GEE, mixed model and fixed-effects methods to data in which newly minted PhDs are clustered within universities. The sample consisted of 554 male biochemists who got their doctorates from 106 American universities in the late 1950s and early 1960s (McGinnis, Allison and Long 1982). The outcome variable PDOC is coded 1 if they got postdoctoral training, and 0 otherwise. The sample is about evenly split, with 49% getting some form of postdoctoral training immediately after receiving their degrees. Other variables examined here are

AGE	Age in years at completion of the PhD
MAR	1 if married, 0 otherwise
DOC	Measure of the prestige of the doctoral institution in bioscience fields
UND	Measure of selectivity of the person's undergraduate institution
AG	1 if degree is from an agricultural department, 0 otherwise
ARTS	Number of articles published while a graduate student
CITS	Number of citations to published articles
DOCID	ID number of the doctoral institution

The first 20 observations in the data set are shown in Output 8.14.

Output 8.14 First 20 Observations in Postdoctoral Data Set

Obs	PDOC	AGE	MAR	DOC	AG	UND	ARTS	CITS
1	0	29	1	362	0	7	1	2
2	1	32	1	210	0	6	1	4
3	1	26	1	359	0	6	0	1
4	0	25	1	181	0	3	0	1
5	1	30	1	429	0	7	0	1
6	1	28	1	359	0	6	1	0
7	1	30	1	210	0	4	0	0
8	1	40	1	347	0	4	0	0
9	0	30	1	210	0	5	1	2
10	1	28	1	359	0	7	5	32
11	0	29	1	447	0	4	0	0
12	1	28	1	276	0	5	1	0
13	0	41	1	261	0	3	0	0
14	1	27	1	226	0	7	0	1
15	1	35	1	359	0	6	0	1
16	1	30	0	341	0	5	7	16
17	0	28	0	226	0	1	1	1
18	1	28	1	429	0	4	3	3
19	0	27	0	359	0	3	0	0
20	1	27	0	205	0	4	0	0

Using PROC LOGISTIC, I first estimated a conventional logit model. The results are shown in Output 8.15. Except for article and citation counts, all variables are statistically significant at the .05 level. Older, married PhDs with degrees from agricultural departments were less likely to receive postdoctoral training. Those from high prestige institutions or who were undergraduates at selective institutions were more likely to receive postdoctoral training.

Output 8.15 PROC LOGISTIC Estimates for Conventional Logit Model

Analysis Of Maximum Likelihood Parameter Estimates							
Parameter	DF	Estimate	Standard Error	Wald 95% Confidence Limits		Wald Chi-Square	Pr > ChiSq
Intercept	1	2.3974	0.9092	0.6154	4.1794	6.95	0.0084
AGE	1	-0.0991	0.0254	-0.1490	-0.0493	15.19	<.0001
MAR	1	-0.5645	0.2684	-1.0905	-0.0385	4.42	0.0354
DOC	1	0.0027	0.0009	0.0009	0.0044	8.40	0.0038
AG	1	-1.1070	0.2030	-1.5049	-0.7090	29.72	<.0001
UND	1	0.1215	0.0600	0.0039	0.2391	4.10	0.0429
ARTS	1	-0.0739	0.0809	-0.2324	0.0846	0.83	0.3610
CITS	1	-0.0030	0.0170	-0.0363	0.0303	0.03	0.8593

Because many of the members in the sample got their degrees from the same institutions, it's reasonable to suspect some dependence among the observations. There are many characteristics of universities besides prestige that are likely to have some impact on the probability of postdoctoral training. The omission of these variables from the model would produce a correlation in postdoctoral training among those men from the same institution. To adjust for this possible correlation, I estimated the model by GEE:

```
PROC GENMOD DATA=postdoc DESC;
  CLASS docid;
  MODEL pdoc = age mar doc ag und arts cits / D=B;
  REPEATED SUBJECT=docid / TYPE=EXCH;
RUN;
```

As we saw in the PTSD example, the TYPE=EXCH option specifies a single correlation that applies to any pair of persons within each cluster. In applications like this one, where there is no natural ordering of persons within clusters, the EXCH option is the only sensible specification.

Output 8.16 GEE Estimates for Postdoctoral Example

Exchangeable Working Correlation	
Correlation	0.0364325063

Analysis Of GEE Parameter Estimates						
Empirical Standard Error Estimates						
Parameter	Estimate	Standard Error	95% Confidence Limits		Z	Pr > |Z|
Intercept	2.4596	0.9468	0.6038	4.3153	2.60	0.0094
AGE	-0.1009	0.0246	-0.1492	-0.0527	-4.10	<.0001
MAR	-0.5611	0.2695	-1.0892	-0.0330	-2.08	0.0373
DOC	0.0027	0.0012	0.0004	0.0050	2.31	0.0210
AG	-0.9930	0.2449	-1.4729	-0.5130	-4.06	<.0001
UND	0.1182	0.0506	0.0191	0.2174	2.34	0.0194
ARTS	-0.0726	0.0802	-0.2298	0.0846	-0.90	0.3656
CITS	-0.0034	0.0168	-0.0363	0.0295	-0.20	0.8395

In Output 8.16, we see that the "exchangeable working correlation" is quite small, only .036. This is the estimated residual correlation between the outcomes for two randomly chosen persons in the same university. Given this small correlation, it's not surprising that the coefficients and standard errors aren't very different from the conventional logistic coefficients in Output 8.15. Table 8.2 converts the z-statistics to chi-squares so that the two methods can be more easily compared. (All chi-squares greater than 3.84 are statistically significant at the .05 level.) Some of the chi-squares are dramatically lower for GEE while others change very little. Those that changed the most—like AG and DOC—are still highly significant.

Table 8.2 Chi-Squares for Different Methods, Postdoctoral Example

	Standard Logit	GEE Robust SE's	Mixed Model Robust SE's	Fixed Effects
AGE	15.19	16.81	16.00	12.00
MAR	4.42	4.33	4.04	2.87
DOC	8.40	5.34	5.11	---
AG	29.72	16.48	12.96	5.98
UND	4.10	5.48	5.71	3.48
ARTS	.83	.81	.76	.79
CITS	.03	.04	.05	.22

Now let's try a random intercept model with PROC GLIMMIX. The SAS code is

```
PROC GLIMMIX DATA=postdoc METHOD=QUAD EMPIRICAL;
  CLASS docid;
  MODEL pdoc = age mar doc ag und arts cits / D=B S;
  RANDOM INTERCEPT / SUBJECT=docid;
  COVTEST 0;
RUN;
```

Notice that I have specified METHOD=QUAD to get true ML estimates, and I've also requested robust standard errors with the EMPIRICAL option. Although that option is not essential, it does reassure us that the standard errors will be appropriate even if the exchangeability assumption is violated.

Results in Output 8.17 are quite similar to the GEE estimates in Output 8.16. Although, as expected, all the coefficients in Output 8.17 are larger than those in Output 8.16, the differences are tiny. The estimated variance of the random intercept (.4036) is much lower than the 2.12 we got for the PTSD data. If we do a Wald test by taking the ratio of the variance to its standard error, we might conclude that there's little evidence that this variance is greater than 0. However, the more definitive likelihood ratio chi-square test at the bottom of the output indicates that this variance is significantly different from 0.

Output 8.17 PROC GLIMMIX Results for Postdoctoral Data

Covariance Parameter Estimates			
Cov Parm	Subject	Estimate	Standard Error
Intercept	DOCID	0.4036	0.3285

Solutions for Fixed Effects					
Effect	Estimate	Standard Error	DF	t Value	Pr > |t|
Intercept	2.6653	1.0305	104	2.59	0.0111
AGE	-0.1101	0.02753	442	-4.00	<.0001
MAR	-0.5870	0.2914	442	-2.01	0.0446
DOC	0.002958	0.001311	442	2.26	0.0245
AG	-1.0215	0.2834	442	-3.60	0.0003
UND	0.1278	0.05349	442	2.39	0.0173
ARTS	-0.07744	0.08917	442	-0.87	0.3857
CITS	-0.00429	0.01868	442	-0.23	0.8186

Tests of Covariance Parameters Based on the Likelihood					
Label	DF	-2 Log Like	ChiSq	Pr > ChiSq	Note
Parameter list	1	688.65	5.25	0.0109	MI

Next, we apply the fixed effects method using PROC LOGISTIC. The attraction of the fixed effects approach is that we control for *all* characteristics of universities, not just prestige. Failure to control for these characteristics could bias some or all of the coefficients in the model.

We implement the method with the program

```
PROC LOGISTIC DATA=postdoc;
  MODEL pdoc(EVENT='1') = age mar ag und arts cits;
  STRATA docid;
RUN;
```

I excluded DOC from the model because it has the same value for all students within each university. If I had put it in, LOGISTIC wouldn't have reported any results for it.

The results in Output 8.18 show little difference in the coefficients as compared with the GEE and mixed model results. Apparently, other characteristics of universities either have little effect on who gets postdoctoral training or are uncorrelated with the measured variables in the model. Nevertheless, the standard errors are higher and the chi-squares are lower than those we got with GEE or the mixed model. That's due in part to the fact that 47 universities with 110 students were excluded from the analysis because either everyone or no one at those universities got postdoctoral training (see the "Model Information" table). Of these universities, 20 were necessarily excluded because the university had only one student in the sample. In any case, the decline in chi-squares is enough to move the *p*-values above the .05 level for two of the variables, MAR and UND. Again, it seems that fixed-effects is not the best method for these data because it produces little reduction in bias but a big increase in standard errors.

Output 8.18 Fixed-Effects Results for Postdoctoral Data

Model Information	
Data Set	WORK.POSTDOC
Response Variable	PDOC
Number of Response Levels	2
Number of Strata	106
Number of Uninformative Strata	47
Frequency Uninformative	110
Model	binary logit
Optimization Technique	Newton-Raphson ridge

Analysis of Maximum Likelihood Estimates					
Parameter	DF	Estimate	Standard Error	Wald Chi-Square	Pr > ChiSq
AGE	1	-0.1076	0.0311	11.9972	0.0005
MAR	1	-0.5314	0.3138	2.8681	0.0903
AG	1	-0.9355	0.3827	5.9757	0.0145
UND	1	0.1352	0.0725	3.4825	0.0620
ARTS	1	-0.0868	0.0976	0.7910	0.3738
CITS	1	-0.0107	0.0228	0.2197	0.6393

8.8 Matching

In the postdoctoral example, the data were clustered into naturally occurring groups. *Matching* is another form of clustering in which individuals are grouped together by design. Matching was once commonly used in the social sciences to control for potentially confounding variables. Now, most researchers use some kind of regression procedure, primarily because of the difficulty of matching on several variables. However, with the development of the *propensity score* method, that objection to matching is largely obsolete (Rosenbaum and Rubin 1983, Smith 1997).

Here's a typical application of the propensity score method. Imagine that your goal is to compare academic achievement of students in public and private schools, controlling for several measures of family background. You have measures on all the relevant variables for a large sample of eighth grade students, 10% of whom are in private schools. The first step in a propensity score analysis is to do a logistic regression in which the dependent variable is the type of school and the independent variables are the family background characteristics. Based on that regression, the propensity score is the predicted probability of being in a private school. Each private school student is then matched to one or more public school students according to their closeness on the propensity score. In most cases, this method produces two groups that have nearly equal means on all the variables in the propensity score regression. One can then do a simple bivariate analysis of achievement versus school type. Alternatively, other variables not in the propensity score regression could be included in some kind of regression analysis. In many situations, this method could have important advantages over conventional regression analysis in reducing both bias and sampling variability (Smith 1997).

The propensity score method is an example of *treatment-control* matching. In this kind of matching, the individuals within each match group necessarily differ on the explanatory variable of central interest. Another sort of matching is *case-control* matching in which individuals within each match group necessarily differ on the *dependent* variable. Case-control studies have long been popular in the biomedical sciences for reasons I explained in Section 3.13. In a case-control study, the aim is to model the determinants of some dichotomous outcome, for instance, a disease condition. People who have the condition are called cases; people who do not have the condition are called controls. In Section 3.13, we saw that it's legitimate to take all the available cases and a random subsample of the controls, pool the two groups into a single sample, and do a conventional logit analysis for the dichotomous outcome. Although not an essential feature of the method, each case is often matched to one or more controls on variables—such as age—that are known to affect the outcome but are not of direct interest.

Although it's usually desirable to adjust for matching in the analysis, the appropriate adjustment methods are quite different for treatment-control and case-control matching. In brief, there are several ways to do it for treatment-control matching, but there is only one way for case-control matching. Let's begin with a treatment-control example. Metraux and Culhane (1999) constructed a data set of 8,402 women who stayed in family shelters in New York City for at least one 7-day period during 1992. The data contained information on several pre-stay characteristics, events that occurred during the stay, and housing type subsequent to the stay. As our dependent variable, we'll focus on whether or not the woman exited to public housing (PUBHOUSE), which was the destination for 48% of the sample. Our principal independent variable is STAYBABY, equal to 1 if a woman gave birth during her stay at the shelter, and 0 otherwise. Nine percent of the sample had a birth during the stay. Other independent variables are

BLACK	1=black race, 0=nonblack
KIDS	Number of children in the household
DOUBLEUP	1=living with another family prior to shelter stay, 0 otherwise
AGE	Age of woman at beginning of shelter stay
DAYS	Number of days in shelter stay

The conventional approach to analysis would be to estimate a logistic regression model for the entire sample of 8,402 women. Results for doing that with PROC LOGISTIC are shown in Output 8.19. We see that although the odds of exiting to public housing increase with the number of children in the household, a birth during the stay reduces the odds by about 33 percent. Both of these effects are overshadowed by the enormous impact of the length of stay. Each additional day increases the odds of exiting to public housing by about 1%. (Not surprisingly, DAYS also has a correlation of .25 with STAYBABY—the longer a woman's stay, the more likely it is that she gave birth during the stay.)

Output 8.19 Logistic Regression of Exit to Public Housing

Analysis of Maximum Likelihood Estimates					
Parameter	**DF**	**Estimate**	**Standard Error**	**Wald Chi-Square**	**Pr > ChiSq**
Intercept	1	-1.7441	0.1543	127.7887	<.0001
STAYBABY	1	-0.4030	0.0990	16.5780	<.0001
BLACK	1	-0.1606	0.0586	7.5134	0.0061
KIDS	1	0.1835	0.0219	70.0082	<.0001
DOUBLEUP	1	-0.1903	0.0601	10.0145	0.0016
AGE	1	-0.0227	0.00523	18.8239	<.0001
DAYS	1	0.0114	0.000257	1973.0471	<.0001

Odds Ratio Estimates			
Effect	**Point Estimate**	**95% Wald Confidence Limits**	
STAYBABY	0.668	0.550	0.811
BLACK	0.852	0.759	0.955
KIDS	1.201	1.151	1.254
DOUBLEUP	0.827	0.735	0.930
AGE	0.978	0.968	0.988
DAYS	1.011	1.011	1.012

Now let's estimate the effect of STAYBABY in a matched sample. I compared all 791 women who had a baby during the stay with an equal number of women who did not. To control for the effect of DAYS, I matched each woman who had a baby with a random draw from among those women whose length of stay was identical (or as close as possible). In this subsample, then, the correlation between STAYBABY and DAYS is necessarily .00. Rosenbaum and Rubin (1983) argue that adjustment by matching is "usually more robust to departures from the assumed form of the underlying model than model-based adjustment on random samples . . . primarily because of reduced reliance on the model's extrapolations." Despite the fact that 6,820 cases are discarded in the matched analysis, we'll see that very little precision is lost in the estimation of the STAYBABY coefficient. I could have used the propensity score method to control for *all* the variables in Output 8.19, but I wanted to keep things simple, and the adjustment for DAYS would have dominated anyway.

Next, I estimated a logistic model for this matched-pair subsample without adjusting for the matching (see Output 8.20). Although the standard errors increase greatly for most of the variables (compared with those in Output 8.19), the loss of precision is quite small for the STAYBABY coefficient; its standard error only increases by about 18% relative to that in Output 8.19. The coefficient for STAYBABY declines somewhat in magnitude, but the matched subsample may be less prone to bias in this estimate than that in the full sample. Note that if we deleted DAYS from this model, the results for STAYBABY would hardly change at all because the two variables are uncorrelated, by design.

Output 8.20 Regression of PUBHOUSE with Matched-Pair Data without Adjustment for Matching

Analysis of Maximum Likelihood Estimates					
Parameter	DF	Estimate	Standard Error	Wald Chi-Square	Pr > ChiSq
Intercept	1	-0.0317	0.3357	0.0089	0.9248
STAYBABY	1	-0.3107	0.1167	7.0914	0.0077
BLACK	1	-0.1322	0.1253	1.1139	0.2912
KIDS	1	0.1789	0.0465	14.8089	0.0001
DOUBLEUP	1	-0.1046	0.1216	0.7397	0.3898
AGE	1	-0.0188	0.0110	2.9045	0.0883
DAYS	1	0.00432	0.000443	95.4934	<.0001

The problem with the analysis in Output 8.20 is that the matched pairs are not independent, which could lead to bias in the standard error estimates. One way to adjust for the matching is to use the GEE method discussed in Section 8.4. Here's how:

```
PROC GENMOD DATA=casecont DESC;
  CLASS casenum;
  MODEL pubhouse=staybaby black kids doubleup age days /
    D=B;
  REPEATED SUBJECT=casenum / TYPE=EXCH;
RUN;
```

CASENUM is the unique identifier for each matched pair. Although I specified an exchangeable structure (TYPE=EXCH), all types of correlation structure are equivalent when there are only two observations per cluster.

Output 8.21 Regression of PUBHOUSE with Matched-Pair Data Using GEE Adjustment for Matching

Exchangeable Working Correlation	
Correlation	0.070850115

Analysis Of GEE Parameter Estimates						
Empirical Standard Error Estimates						
Parameter	Estimate	Standard Error	95% Confidence Limits		Z	Pr > |Z|
Intercept	-0.0124	0.3487	-0.6959	0.6710	-0.04	0.9716
STAYBABY	-0.3100	0.1074	-0.5206	-0.0995	-2.89	0.0039
BLACK	-0.1462	0.1197	-0.3809	0.0884	-1.22	0.2218
KIDS	0.1790	0.0469	0.0871	0.2709	3.82	0.0001
DOUBLEUP	-0.1138	0.1168	-0.3427	0.1151	-0.97	0.3298
AGE	-0.0188	0.0108	-0.0400	0.0024	-1.74	0.0819
DAYS	0.0043	0.0007	0.0030	0.0056	6.52	<.0001

Results in Output 8.21 differ only slightly from those in Output 8.20, which were not adjusted for matching. In particular, the coefficient for STAYBABY is about the same, while its estimated standard error is slightly reduced, from .1167 to .1074. Consistent with the small change is the estimated "residual" correlation between within-pair observations of only .07. One reason why the results don't differ more is that the inclusion of DAYS in Output 8.20 is itself a partial adjustment for matching. When DAYS is omitted from the model, the residual correlation is .20.

In my opinion, GEE is usually the best method for adjusting for treatment-control matching. Although the example consisted of matched pairs, the method is identical for one-to-many matching or many-to-many matching. One alternative to GEE is the mixed model discussed in Section 8.5, but this is likely to give very similar results in most applications. Another widely recommended alternative is the conditional logit (fixed effects) model of Section 8.6. Unfortunately, this method often involves a substantial loss of data with concomitant increases in standard errors. The shelter stay data provides a good example of this loss. The LOGISTIC program for the fixed-effects model with matched-pair data is

```
PROC LOGISTIC DATA=casecont;
  MODEL pubhouse(EVENT='1')= staybaby black kids doubleup age;
  STRATA casenum;
RUN;
```

There is no need to include DAYS in the model because it is the same (or almost the same) for both members of every matched pair.

Output 8.22 PROC LOGISTIC Output for Fixed Effects Analysis of Matched Pairs

Analysis of Maximum Likelihood Estimates					
Parameter	DF	Estimate	Standard Error	Wald Chi-Square	Pr > ChiSq
STAYBABY	1	-0.3850	0.1367	7.9370	0.0048
BLACK	1	-0.3275	0.1939	2.8547	0.0911
KIDS	1	0.2015	0.0714	7.9658	0.0048
DOUBLEUP	1	-0.3283	0.2043	2.5827	0.1080
AGE	1	-0.0283	0.0181	2.4584	0.1169

The coefficients in Output 8.22 are quite similar to those in Output 8.21, but the standard error of STAYBABY is about 27% higher than it was before. The increase is attributable to the fact that the conditional logit method discarded 525 out of the 791 matched pairs because both members of the pair had the same value of the dependent variable PUBHOUSE. In Section 8.6, I argued that this loss of information must be balanced by the potential decrease in bias that comes from controlling all stable characteristics of the cluster that might be correlated with the treatment variable, in this case, STAYBABY. However, because the matching is balanced in this application (with one treatment and one control per cluster), it's impossible for STAYBABY to be correlated with cluster characteristics. So, there's no potential benefit from the conditional logit method.

The situation is quite different for case-control designs. In that setting, every matched pair has one each of the two values of the *dependent* variable. If you try to apply GEE to such data, the estimated working residual correlation is -1 and the method breaks down. A random intercept model is even worse because it can only accommodate a positive correlation. On the other hand, the conditional logit method suffers no loss of data because there are no clusters in which both members have the same value on the dependent variable. So, conditional logit is the only way to go for case-control matching.

As an example of the case-control design, we again use the data on shelter stays, but now with STAYBABY as the dependent variable rather than as an independent variable. Output 8.23 shows the results from estimating a logistic model (with PROC LOGISTIC) for the full data set of 8,402 women. We see evidence that the probability of having a baby during the shelter stay is higher for blacks, women with more children in the household, younger women, and those with longer shelter stays.

Output 8.23 Logistic Regression of STAYBABY for Full Sample

Response Profile		
Ordered Value	STAYBABY	Total Frequency
1	0	7611
2	1	791

Analysis of Maximum Likelihood Estimates					
Parameter	DF	Estimate	Standard Error	Wald Chi-Square	Pr > ChiSq
Intercept	1	-2.6802	0.2204	147.8532	<.0001
BLACK	1	0.2857	0.0860	11.0466	0.0009
KIDS	1	0.1558	0.0287	29.5256	<.0001
DOUBLEUP	1	0.1497	0.0832	3.2351	0.0721
AGE	1	-0.0467	0.00772	36.6776	<.0001
DAYS	1	0.00447	0.000224	399.7093	<.0001

Now let's re-estimate the model for the 791 pairs of women who are matched by number of days of shelter stay. The LOGISTIC output without adjustment for matching is shown in Output 8.24.

Output 8.24 Logistic Regression of STAYBABY for Matched Pairs, No Adjustment for Matching

Response Profile		
Ordered Value	STAYBABY	Total Frequency
1	0	791
2	1	791

Analysis of Maximum Likelihood Estimates					
Parameter	DF	Estimate	Standard Error	Wald Chi-Square	Pr > ChiSq
Intercept	1	0.6315	0.2870	4.8413	0.0278
BLACK	1	0.3435	0.1101	9.7308	0.0018
KIDS	1	0.1717	0.0386	19.8111	<.0001
DOUBLEUP	1	0.1638	0.1094	2.2426	0.1343
AGE	1	-0.0464	0.00977	22.5137	<.0001
DAYS	1	-0.00004	0.000300	0.0212	0.8842

I included DAYS in the model just to demonstrate that matching on this variable eliminates its effect on the dependent variable. The other variables have coefficient estimates that are quite similar to those for the full sample. The standard errors are about 30% larger than in the full sample, but that's not bad considering that we have discarded 81% of the cases.

We still haven't adjusted for matching, however. To do that, we include a STRATA statement in the LOGISTIC program:

```
PROC LOGISTIC DATA=casecont;
  MODEL staybaby(EVENT='1')=black kids doubleup age;
  STRATA casenum;
RUN;
```

The results in Output 8.25 are very close to those in Output 8.24, which did not adjust for matching.

Output 8.25 Logistic Regression of STAYBABY for Matched Pairs, Adjusted for Matching

Analysis of Maximum Likelihood Estimates					
Parameter	DF	Estimate	Standard Error	Wald Chi-Square	Pr > ChiSq
BLACK	1	0.3535	0.1092	10.4763	0.0012
KIDS	1	0.1866	0.0399	21.8291	<.0001
DOUBLEUP	1	0.1865	0.1114	2.8020	0.0941
AGE	1	-0.0496	0.0102	23.7281	<.0001

8.9 Comparison of Methods

Table 8.4 summarizes differences among the four methods for handling clustered data. All four methods produce consistent estimates of the standard errors of the coefficients in the presence of clustering. But GEE goes further than robust variance estimation because it also produces coefficient estimates that have minimal sampling variability. On the other hand, GEE gives estimates of population-averaged coefficients rather than subject-specific coefficients. Population-averaged coefficients are subject to heterogeneity shrinkage—attenuation toward 0 in the presence of heterogeneity in the population. Heterogeneity shrinkage can be corrected by estimating either a conditional logistic regression or a mixed model. Conditional logit goes further by reducing bias that arises from correlations between individual and cluster-level variables (spuriousness). But conditional logit analysis may also discard a considerable portion of the data, thereby increasing the standard errors.

Table 8.4 Effectiveness of Different Methods for Logit Analysis of Clustered Data

	Coefficient Bias			
	Corrects Spuriousness	Corrects Shrinkage	Corrects S.E. Bias	Efficiency
Conditional Logit	✔	✔	✔	
Mixed Models		✔	✔	✔
GEE			✔	✔
Robust S.E.'s			✔	

The best method to use depends greatly on the design, the data, and the goals of the analysis. In randomized experiments, there is little danger of spuriousness, so conditional logit is relatively unattractive. With observational (non-experimental) data, on the other hand, spuriousness is nearly always a serious concern, making conditional logit analysis especially attractive. I typically use both conditional logit and GEE analysis (or a mixed model) to get greater insight into the data. But as we saw in the matching section, some designs make GEE impossible while others make conditional logit analysis unappealing.

8.10 A Hybrid Method

The four methods just considered are now fairly well known and understood. There's also a fifth approach that combines the virtues of the conditional logit method with those of GEE or the mixed model. Neuhaus and Kalbfleisch (1998) have the most explicit discussion of the method for binary data, but the general strategy was previously considered by Bryk and Raudenbusch (1992) and applied to a study of criminal recidivism by Horney et al. (1995). The method can be summarized in four steps:

1. Calculate the means of the time-varying explanatory variables for each individual.
2. Calculate the deviations of the time-varying explanatory variables from the individual-specific means.
3. Estimate the model with variables created in steps 1 and 2, along with any additional time-constant explanatory variables.
4. Use GEE or a mixed model to adjust for residual dependence.

The coefficients and standard errors for the deviation variables are typically very similar to those obtained with conditional logit analysis. But unlike conditional logit analysis, this method also allows for the inclusion of explanatory variables that are constant over time.

Before estimating the model, one must calculate the individual-specific means and the deviations from those means. Here's how to do it for the PTSD data:

```
PROC SUMMARY DATA=ptsd NWAY;
  CLASS subjid;
  VAR control problems sevent;
  OUTPUT OUT=means MEAN=mcontrol mproblem msevent;
DATA combine;
  MERGE ptsd means;
  BY subjid;
```

```
  dcontrol=control-mcontrol;
  dproblem=problems-mproblem;
  dsevent=sevent-msevent;
RUN;
```

PROC SUMMARY produces a data set called MEANS, which contains 316 observations. Each of these observations has four variables: the individual's ID number and the individual's means for the three time-varying explanatory variables (CONTROL, PROBLEMS, and SEVENT). The DATA step merges the MEANS data set with the original data set and computes the deviation variables. The new data set (COMBINE) has 948 observations, with the means replicated across the three observations for each individual.

We can now estimate a logistic model that includes both the mean and the deviations from the means. Because there may still be residual dependence in the dependent variable, this should be accommodated with either GEE or a mixed model. The SAS code for a mixed model using PROC GLIMMIX is:

```
PROC GLIMMIX DATA=combine METHOD=QUAD EMPIRICAL;
  CLASS subjid time;
  MODEL ptsd = dcontrol dproblem dsevent mcontrol mproblem
       msevent cohes time / D=B S;
  RANDOM INTERCEPT / SUBJECT=subjid;
RUN;
```

I've included the EMPIRICAL option to produce robust standard errors which protect against violation of the exchangeability assumption that is implicit in the random intercept model. The results are shown in Output 8.26. As expected, the coefficients and *p*-values for the three deviation variables are quite close to those in Output 8.13 for the conditional logit model. In this example, coefficients for the three mean variables are also somewhat similar to those for the deviation variables, but this will not always be the case. In fact, I've seen examples in which the deviation variables and the mean variables have highly significant coefficients that are opposite in sign.

Output 8.26 PROC GLIMMIX Estimates for Hybrid PTSD Model

Covariance Parameter Estimates			
Cov Parm	Subject	Estimate	Standard Error
Intercept	subjid	2.1220	0.6115

Solutions for Fixed Effects						
Effect	time	Estimate	Standard Error	DF	t Value	Pr > |t|
Intercept		0.4516	1.7457	312	0.26	0.7961
dcontrol		-1.1815	0.4131	626	-2.86	0.0044
dproblem		0.2009	0.09911	626	2.03	0.0431
dsevent		0.1946	0.1538	626	1.27	0.2062
mcontrol		-1.1438	0.4210	626	-2.72	0.0068
mproblem		0.4913	0.09908	626	4.96	<.0001
msevent		0.8033	0.1966	626	4.09	<.0001
cohes		-0.2345	0.06373	626	-3.68	0.0003
time	1	0.8433	0.2873	626	2.94	0.0035
time	2	0.4433	0.2324	626	1.91	0.0569
time	3	0	.	.	.	.

While the deviation coefficients can be interpreted in the same way as the conditional logit coefficients, the coefficients for the mean variables are usually difficult to interpret causally. Under the assumptions of the conventional random intercept model of Section 8.5, the coefficients for the mean variables and the corresponding deviation variables should be identical. Hence, the degree to which they differ can be the basis of a specification test for the conventional model. We can test the null hypothesis that the coefficients are the same by including a CONTRAST statement in the hybrid model program:

```
CONTRAST 'spec test' dcontrol 1 mcontrol -1,
     dproblem 1 mproblem -1, dsevent 1 msevent -1;
```

This produced an *F*-statistic of 4.25 with 3 and 626 degrees of freedom, yielding a *p*-value of .0055. Apparently, at least one pair of coefficients is different, suggesting that the conventional mixed model could lead to biased estimates of the coefficients.

Although the hybrid method seems to produce useful results that are quite similar to those of the conditional logit model, its statistical properties have not been thoroughly investigated. In particular, I have not seen any proof of its statistical consistency or efficiency.

Chapter 9

Regression for Count Data

9.1 Introduction

In this chapter, we examine Poisson regression and negative binomial regression, which are two methods that are appropriate for dependent variables that have only non-negative integer values: 0, 1, 2, 3, etc. Usually these numbers represent counts of something, like number of people in an organization, number of visits to a physician, or number of arrests in the past year. While such data are fairly common in the social sciences, there is another reason why Poisson regression is important: it is a fundamental building block for loglinear analysis of contingency tables, which is a topic we will cover in some detail in the next chapter.

For years, people analyzed count data by ordinary linear regression and, for many applications, that method was adequate for the task. But Poisson and negative binomial regression have the advantage of being precisely tailored to the discrete, often highly skewed distribution of the dependent variable. On the downside, Poisson regression has the disadvantage of being susceptible to problems of overdispersion that do not affect ordinary linear regression. Overdispersion, discussed in detail later, can produce severe underestimates of standard errors and overestimates of test statistics. While there are some simple corrections for overdispersion, negative binomial regression is generally the preferred method whenever there is evidence for overdispersion.

9.2 The Poisson Regression Model

The Poisson regression model gets its name from the assumption that the dependent variable has a Poisson distribution, defined as follows. Let y be a variable that can have only non-negative integer values. We assume that the probability that y is equal to some number r is given by

$$\Pr(y=r)=\frac{\lambda^r e^{-\lambda}}{r!}, \qquad r=0,1,2,\ldots \tag{9.1}$$

where λ is the expected value (mean) of y and $r!=r(r-1)(r-2)...(1)$. Although y can only take on integer values, λ can be any positive number. For λ=1.5, the probabilities for the Poisson distribution are graphed in Figure 9.1.

Figure 9.1 Poisson Distribution for λ = 1.5

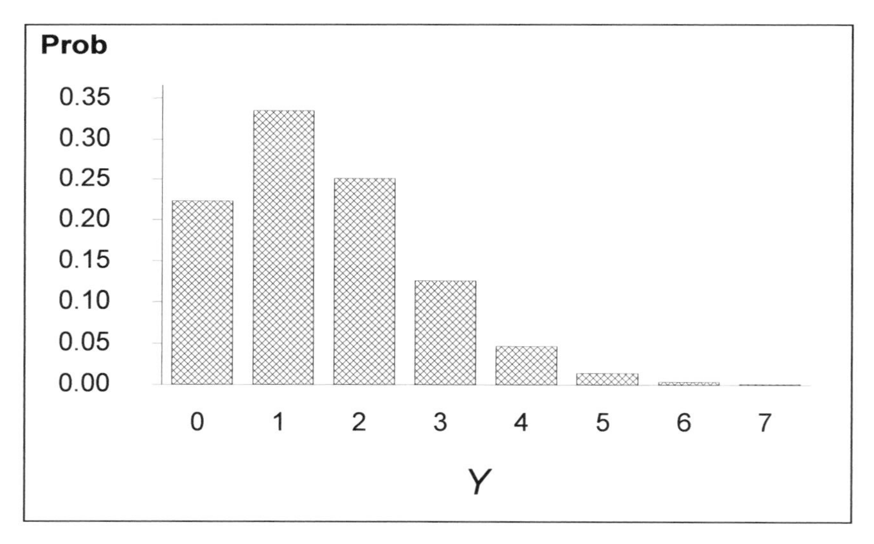

As λ gets larger, the mode moves away from 0 and the distribution looks more and more like a normal distribution. An unusual property of the Poisson distribution is that the mean and variance are equal,

$$\text{E}(y) = \text{var}(y) = \lambda \tag{9.2}$$

which turns out to be problematic in many applications.

Next we need to specify how the parameter λ depends on the explanatory variables. First, we write λ_i with a subscript *i* to allow the parameter to vary across individuals (i= 1, …, n). Then, because λ can't be less than 0, it is standard to let λ be a loglinear function of the *x* variables:

$$\log \lambda_i = \beta_0 + \beta_1 x_{i1} + \beta_2 x_{i2} + \ldots + \beta_k x_{ik}. \tag{9.3}$$

This ensures that λ will be greater than 0 for any values of the *x*'s or the β's.

That's all there is to the model. Note that the model does not say that the marginal distribution of *y* will necessarily be Poisson. (The marginal distribution is what you get when you ignore any explanatory variables. Outputs 9.1 and 9.2 are realizations of marginal distributions.) Instead, *y* has a Poisson distribution *conditional* on the values of the explanatory variables. If the *x* variables have large coefficients and large variances, the marginal distribution of *y* may look very different from a Poisson distribution.

As usual, we'll estimate the model by maximum likelihood. This is easily accomplished with PROC GENMOD.

9.3 Scientific Productivity Example

In Chapter 8, we studied a sample of 554 biochemists who got their doctorates in the late 1950s and early 1960s. Two of the variables in the data set are good candidates for Poisson regression: ARTS—the number of articles published by the biochemists before they received their degrees (counted in *Chemical Abstracts*), and CITS—the number of citations to those articles (counted in *Science Citation Index*). Output 9.1 and Output 9.2 show the marginal frequency distributions for these two variables.

Output 9.1 Frequency Distribution of Article Counts

ARTS	Frequency	Percent	Cumulative Frequency	Cumulative Percent
0	330	59.57	330	59.57
1	110	19.86	440	79.42
2	57	10.29	497	89.71
3	33	5.96	530	95.67
4	9	1.62	539	97.29
5	5	0.90	544	98.19
6	3	0.54	547	98.74
7	3	0.54	550	99.28
10	2	0.36	552	99.64
11	1	0.18	553	99.82
15	1	0.18	554	100.00

Both distributions are highly skewed, especially citation counts. Neither distribution would pass a statistical test for the Poisson distribution, but as I noted earlier, that's not essential for the model to be correct.

Output 9.2 Frequency Distribution of Citation Counts

CITS	Frequency	Percent	Cumulative Frequency	Cumulative Percent
0	187	33.75	187	33.75
1	188	33.94	375	67.69
2	36	6.50	411	74.19
3	25	4.51	436	78.70
4	12	2.17	448	80.87
5	23	4.15	471	85.02
6	5	0.90	476	85.92
7	19	3.43	495	89.35
8	4	0.72	499	90.07
9	4	0.72	503	90.79
10	7	1.26	510	92.06
11	5	0.90	515	92.96
12	3	0.54	518	93.50
13	1	0.18	519	93.68
14	1	0.18	520	93.86
15	5	0.90	525	94.77
16	4	0.72	529	95.49
17	1	0.18	530	95.67
20	5	0.90	535	96.57
22	2	0.36	537	96.93
23	3	0.54	540	97.47
25	1	0.18	541	97.65
27	2	0.36	543	98.01
30	1	0.18	544	98.19
32	3	0.54	547	98.74
33	1	0.18	548	98.92
37	3	0.54	551	99.46
40	1	0.18	552	99.64
57	1	0.18	553	99.82
74	1	0.18	554	100.00

The next step is to estimate Poisson regressions for articles and citations as predicted by the other independent variables described in Section 8.5. Here's the GENMOD code for the two regressions:

```
PROC GENMOD DATA=postdoc;
  MODEL arts = age mar doc ag und / DIST=POISSON;
PROC GENMOD DATA=postdoc;
  MODEL cits = age mar doc ag und / D=P;
RUN;
```

To fit a Poisson regression, we use the DIST=POISSON option, which can be abbreviated D=P. When a Poisson regression is requested, the loglinear model in equation (9.3) is the default. Results appear in Output 9.3 and Output 9.4.

Output 9.3 Poisson Regression for Article Counts

Model Information	
Data Set	POSTDOC
Distribution	Poisson
Link Function	Log
Dependent Variable	ARTS

Number of Observations Read	554
Number of Observations Used	554

Criteria For Assessing Goodness Of Fit			
Criterion	DF	Value	Value/DF
Deviance	548	1072.2986	1.9567
Scaled Deviance	548	1072.2986	1.9567
Pearson Chi-Square	548	1474.6054	2.6909
Scaled Pearson X2	548	1474.6054	2.6909
Log Likelihood		-541.0603	
Full Log Likelihood		-813.1613	
AIC (smaller is better)		1638.3226	
AICC (smaller is better)		1638.4762	
BIC (smaller is better)		1664.2256	

Analysis Of Maximum Likelihood Parameter Estimates							
Parameter	DF	Estimate	Standard Error	Wald 95% Confidence Limits		Wald Chi-Square	Pr > ChiSq
Intercept	1	0.4241	0.4645	-0.4862	1.3344	0.83	0.3612
age	1	-0.0312	0.0131	-0.0569	-0.0056	5.70	0.0170
mar	1	0.0087	0.1306	-0.2474	0.2648	0.00	0.9470
doc	1	-0.0001	0.0005	-0.0010	0.0008	0.07	0.7861

Analysis Of Maximum Likelihood Parameter Estimates							
Parameter	DF	Estimate	Standard Error	Wald 95% Confidence Limits		Wald Chi-Square	Pr > ChiSq
ag	1	0.0319	0.1001	-0.1642	0.2281	0.10	0.7495
und	1	0.0698	0.0303	0.0103	0.1292	5.29	0.0214
Scale	0	1.0000	0.0000	1.0000	1.0000		

In Output 9.3, we see that two variables are significant predictors of publication counts, age and the selectivity of the undergraduate institution. Because the dependent variable is logged, we can interpret the coefficients much like logistic regression coefficients. Specifically, if we calculate $100(e^{\beta}-1)$, we get the percent change in the expected number of publications with each 1-unit increase in the independent variable. For AGE, a one-year increase in age is associated with a 3.1% decrease in the expected number of publications. For the 7-point UND scale, the increase in the expected number of publications for each 1-point increase is about 7.2%.

Note that the deviance is almost twice as large as the degrees of freedom in Output 9.3. It's probably not appropriate to calculate a p-value for this statistic because the predicted number of articles is quite small for many of the biochemists. As we saw in fitting logistic regression models with individual level data, when predicted values are small, the deviance is not well approximated by a chi-square distribution. Nevertheless, the large ratio of deviance to degrees of freedom does suggest a problem with the model, one that we'll pursue in the next section. The problem is even more severe in Output 9.4 for citation counts, where the deviance is more than seven times the degrees of freedom. In this regression, all the variables are highly significant. Older students at agricultural schools have lower average citation counts, while married students at prestigious institutions and from selective undergraduate institutions have higher average citation counts. More precisely, the expected citation count for those in agricultural schools is 36% lower than for those in medical schools (calculated as 100(exp(-.4469)-1)=-36). Married students have citation counts that are 100(exp(.29)-1)=34% higher, on average, than those for unmarried students. Before we take these results too seriously, however, we need to consider the problem of overdispersion, exemplified by the large deviance relative to the degrees of freedom.

There's one other thing about these two output examples that merits a few lines of explanation. The "Criteria For Assessing Goodness Of Fit" tables report both a "log likelihood" and a "full log likelihood." What's the difference? The "full log likelihood" is the true logarithm of the likelihood function. The BIC, AIC, and AICC statistics are based on this number. In earlier releases of GENMOD, however, only the "log likelihood" was reported. This statistic differs from the "full log likelihood" by a constant that depends on the data but not on the unknown parameters, and therefore it is sufficient to find parameter values that maximize the "log likelihood." While the "log likelihood" may be positive due to the omission of the unnecessary constant, the "full log likelihood" is always negative.

Output 9.4 Poisson Regression for Citation Counts

Criteria For Assessing Goodness Of Fit			
Criterion	**DF**	**Value**	**Value/DF**
Deviance	548	4053.1740	7.3963
Scaled Deviance	548	4053.1740	7.3963
Pearson Chi-Square	548	7174.8868	13.0929
Scaled Pearson X2	548	7174.8868	13.0929
Log Likelihood		448.8995	
Full Log Likelihood		-2541.4601	
AIC (smaller is better)		5094.9202	
AICC (smaller is better)		5095.0737	
BIC (smaller is better)		5120.8232	

Analysis Of Maximum Likelihood Parameter Estimates							
Parameter	**DF**	**Estimate**	**Standard Error**	**Wald 95% Confidence Limits**		**Wald Chi-Square**	**Pr > ChiSq**
Intercept	1	1.3407	0.2427	0.8651	1.8163	30.53	<.0001
age	1	-0.0417	0.0069	-0.0551	-0.0282	36.73	<.0001
mar	1	0.2899	0.0699	0.1529	0.4269	17.20	<.0001
doc	1	0.0009	0.0002	0.0005	0.0014	15.50	<.0001
ag	1	-0.4469	0.0558	-0.5563	-0.3374	64.03	<.0001
und	1	0.1269	0.0156	0.0963	0.1575	66.03	<.0001
Scale	0	1.0000	0.0000	1.0000	1.0000		

9.4 Overdispersion

I now claim that the standard errors, chi-squares, and *p*-values in the previous section are essentially worthless because of overdispersion. We encountered overdispersion with grouped-data logit models in Chapter 4 where I argued that the problem was primarily restricted to clustered data. For Poisson regression, it's *always* a potential problem, and often quite a serious one.

Basically the problem arises because equation (9.2) says that, for a given set of values on the explanatory variables, the variance of the dependent variable is equal to its mean. In fact, the variance is often much higher than that. Equivalently, we can say that overdispersion occurs because there's no random disturbance term in equation (9.3) that would allow for omitted explanatory variables. (A disturbance term would produce a larger variance in y.) While overdispersion doesn't bias the coefficients, it does lead to underestimates of the standard errors and overestimates of chi-square statistics. Overdispersion also implies that conventional maximum likelihood estimates are not efficient, meaning that other methods can produce coefficients with less sampling variability.

What can be done about overdispersion? If you're willing to ignore the lack of efficiency of conventional estimates, it's a simple matter to correct the standard errors and chi-squares (Agresti 2002). The correction is the same as for the grouped binomial case that we discussed in Chapter 4: take the ratio of the goodness-of-fit chi-square to its degrees of freedom, and call the result C. Divide the chi-square statistic for each coefficient by C. Multiply the standard error of each coefficient by the square root of C. The only ambiguity is that we have two goodness-of-fit chi-squares, the deviance and the Pearson chi-square. Which one should we use? Most of the time they'll be fairly close, but the theory of quasi-likelihood estimation suggests the use of the Pearson chi-square (McCullagh and Nelder 1989).

In GENMOD, the corrections just described can be automatically invoked by putting either SCALE=P (for Pearson) or SCALE=D (for deviance) as options in the MODEL statement. For our biochemist example, the full code using the SCALE=P option is

```
PROC GENMOD DATA=postdoc;
  MODEL arts = age mar doc ag und / D=P SCALE=P;
PROC GENMOD DATA=postdoc;
  MODEL cits = age mar doc ag und / D=P SCALE=P;
RUN;
```

Results are shown in Output 9.5 and Output 9.6. Each output now includes a scale parameter which, as noted, is just the square root of the Pearson chi-square divided by the degrees of freedom. For articles, the chi-squares for the coefficients are all reduced by about 40%, with the result that none of the coefficients is statistically significant. For citations, the chi-square adjustments are much more severe, with reductions of over 70%. The only significant coefficients are for agricultural school and for undergraduate selectivity. Clearly, reliance on the unadjusted test statistics would have been disastrous.

Output 9.5 Poisson Regression for Articles with Overdispersion Correction

Analysis Of Maximum Likelihood Parameter Estimates							
Parameter	DF	Estimate	Standard Error	Wald 95% Confidence Limits		Wald Chi-Square	Pr > ChiSq
Intercept	1	0.4241	0.7619	-1.0692	1.9173	0.31	0.5778
age	1	-0.0312	0.0215	-0.0733	0.0108	2.12	0.1457
mar	1	0.0087	0.2143	-0.4113	0.4287	0.00	0.9677
doc	1	-0.0001	0.0007	-0.0016	0.0013	0.03	0.8686
ag	1	0.0319	0.1641	-0.2898	0.3537	0.04	0.8457
und	1	0.0698	0.0497	-0.0277	0.1673	1.97	0.1607
Scale	0	1.6404	0.0000	1.6404	1.6404		

Note: The scale parameter was estimated by the square root of Pearson's Chi-Square/DOF.

Output 9.6 Poisson Regression for Citations with Overdispersion Correction

Analysis Of Maximum Likelihood Parameter Estimates							
Parameter	DF	Estimate	Standard Error	Wald 95% Confidence Limits		Wald Chi-Square	Pr > ChiSq
Intercept	1	1.3407	0.8781	-0.3803	3.0617	2.33	0.1268
age	1	-0.0417	0.0249	-0.0904	0.0071	2.81	0.0940
mar	1	0.2899	0.2530	-0.2059	0.7857	1.31	0.2518
doc	1	0.0009	0.0009	-0.0007	0.0026	1.18	0.2766
ag	1	-0.4469	0.2021	-0.8429	-0.0508	4.89	0.0270
und	1	0.1269	0.0565	0.0161	0.2376	5.04	0.0247
Scale	0	3.6184	0.0000	3.6184	3.6184		

Note: The scale parameter was estimated by the square root of Pearson's Chi-Square/DOF.

Somewhat surprisingly, ordinary linear regression is not susceptible to the problem of overdispersion because it automatically estimates a scale parameter that is used in calculating standard errors and test statistics. The scale parameter for a linear regression is just the estimated standard deviation of the disturbance term, sometimes called the root mean squared error. To illustrate this point, I used ordinary least squares (OLS) to regress the logarithm of citation counts on the same explanatory variables. Because the logarithm of 0 is undefined, I first added .5 to everyone's citation count:

```
DATA postdoc2;
  SET postdoc;
  lcits=log(cits+.5);
PROC GENMOD DATA=postdoc2;
  MODEL lcits = age mar doc ag und;
RUN;
```

When no DIST option is specified in the MODEL statement, GENMOD uses OLS to estimate a normal linear model. The results shown in Output 9.7, while not identical to those in Output 9.6, are at least in the same ballpark. In general, Poisson regression with a correction for overdispersion is preferable to ordinary least squares. But OLS may be better than Poisson regression without the overdispersion correction.

Output 9.7 OLS Results for Log-Linear Citation Model

Analysis Of Maximum Likelihood Parameter Estimates							
Parameter	DF	Estimate	Standard Error	Wald 95% Confidence Limits		Wald Chi-Square	Pr > ChiSq
Intercept	1	0.6403	0.4647	-0.2705	1.5511	1.90	0.1683
AGE	1	-0.0193	0.0127	-0.0441	0.0055	2.32	0.1278
MAR	1	0.0873	0.1382	-0.1836	0.3581	0.40	0.5277
DOC	1	0.0005	0.0005	-0.0004	0.0015	1.33	0.2489
AG	1	-0.2765	0.1034	-0.4791	-0.0739	7.15	0.0075
und	1	0.0597	0.0311	-0.0012	0.1206	3.69	0.0547
Scale	1	1.1279	0.0339	1.0634	1.1963		

Note: The scale parameter was estimated by maximum likelihood.

9.5 Negative Binomial Regression

The adjustment for overdispersion discussed in the last section is a huge improvement over conventional Poisson regression, but it's not ideal. The coefficients are still inefficient, meaning that they have more sampling variability than necessary. Efficient estimates may be produced by a method known as negative binomial regression that has become increasingly popular for count data.

The negative binomial model is a generalization of the Poisson model. We modify equation (9.3) to include a disturbance term, which accounts for the overdispersion:

$$\log \lambda_i = \beta_0 + \beta_1 x_{i1} + \beta_2 x_{i2} + \ldots + \beta_k x_{ik} + \sigma\varepsilon_i .$$

We assume that the dependent variable y_i has a Poisson distribution with expected value λ_i, *conditional on* ε_i. Finally, we assume that $\exp(\varepsilon_i)$ has a standard gamma distribution (Agresti 2002). It follows that the unconditional distribution of y_i is a negative binomial distribution.

The negative binomial regression model may be efficiently estimated by maximum likelihood. In PROC GENMOD, this is accomplished simply by using the option D=NB on the MODEL statement:

```
PROC GENMOD DATA=postdoc;
  MODEL arts = age mar doc ag und / D=NB;
PROC GENMOD DATA=postdoc;
  MODEL cits = age mar doc ag und / D=NB;
RUN;
```

Results are shown in Output 9.8 and Output 9.9. For articles, the results are quite close to those we got with the simpler adjustment in Output 9.5. Now the deviance is actually less than the degrees of freedom, indicating a very good fit. The dispersion parameter, which is what allows for overdispersion, is estimated to be 1.74. If this were 0, we would be back to the Poisson model. You can get a test for whether the dispersion parameter is 0 by putting the option NOSCALE on the MODEL statement. This constrains the dispersion parameter to be 0 and reports a Lagrange multiplier test for that constraint. For this example, we get a chi-square of 61.5 with one degree of freedom, which is highly significant. So we reject the simpler Poisson model in favor of the (slightly) more complicated negative binomial.

For citations, the differences between the negative binomial model of Output 9.9 and the SCALE=P adjustment of Output 9.6 are more substantial. In Output 9.9, the *p*-values for the significant variables are noticeably lower, and AGE is now significant at about the .01 level. As with publications, the negative binomial model for citations fits very well, as indicated by the fact that the deviance is only slightly larger than the degrees of freedom. Not surprisingly, the dispersion parameter for the citations is substantially larger than the estimate for publications.

Output 9.8 Negative Binomial Regression for Article Counts

Criteria For Assessing Goodness Of Fit			
Criterion	DF	Value	Value/DF
Deviance	548	478.4304	0.8730
Scaled Deviance	548	478.4304	0.8730
Pearson Chi-Square	548	591.6643	1.0797
Scaled Pearson X2	548	591.6643	1.0797
Log Likelihood		-422.8414	
Full Log Likelihood		-694.9425	
AIC (smaller is better)		1403.8849	
AICC (smaller is better)		1404.0900	
BIC (smaller is better)		1434.1051	

Analysis Of Maximum Likelihood Parameter Estimates							
Parameter	DF	Estimate	Standard Error	Wald 95% Confidence Limits		Wald Chi-Square	Pr > ChiSq
Intercept	1	0.2653	0.6594	-1.0270	1.5577	0.16	0.6874
age	1	-0.0278	0.0185	-0.0640	0.0083	2.28	0.1313
mar	1	0.0162	0.2100	-0.3954	0.4278	0.01	0.9384
doc	1	0.0000	0.0007	-0.0013	0.0014	0.00	0.9535
ag	1	0.0371	0.1582	-0.2728	0.3471	0.06	0.8143
und	1	0.0694	0.0466	-0.0219	0.1607	2.22	0.1365
Dispersion	1	1.7418	0.2356	1.3362	2.2704		

Note: The negative binomial dispersion parameter was estimated by maximum likelihood.

Output 9.9 Negative Binomial Regression for Citation Counts

Criteria For Assessing Goodness Of Fit			
Criterion	DF	Value	Value/DF
Deviance	548	561.8670	1.0253
Scaled Deviance	548	561.8670	1.0253
Pearson Chi-Square	548	846.1709	1.5441
Scaled Pearson X2	548	846.1709	1.5441
Log Likelihood		1794.0857	
Full Log Likelihood		-1196.2739	
AIC (smaller is better)		2406.5479	
AICC (smaller is better)		2406.7530	
BIC (smaller is better)		2436.7680	

Analysis Of Maximum Likelihood Parameter Estimates							
Parameter	DF	Estimate	Standard Error	Wald 95% Confidence Limits		Wald Chi-Square	Pr > ChiSq
Intercept	1	1.5560	0.6560	0.2702	2.8418	5.63	0.0177
age	1	-0.0444	0.0178	-0.0792	-0.0096	6.24	0.0125
mar	1	0.2301	0.1968	-0.1557	0.6158	1.37	0.2425
doc	1	0.0010	0.0006	-0.0003	0.0022	2.40	0.1213
ag	1	-0.4598	0.1530	-0.7597	-0.1600	9.04	0.0026
und	1	0.1086	0.0455	0.0195	0.1978	5.71	0.0169
Dispersion	1	2.2597	0.1682	1.9530	2.6146		

Note: The negative binomial dispersion parameter was estimated by maximum likelihood.

Given the known superiority of the negative binomial model to the Poisson model for cases of overdispersion, and given the frequency with which overdispersion occurs, why not always use the negative binomial model? It used to be the case that the negative binomial model was either unavailable or difficult to implement. But now that it's readily available and easy to implement, I don't see any compelling reason for using Poisson regression.

9.6 Adjustment for Varying Time Spans

The Poisson distribution (or its generalization to the negative binomial) is well suited for describing counts of events that occur in some interval of time. In the preceding example, articles and citations were counted over a certain period of time. Other examples include the number of arrests in a five-year period, the number of colds that occur in one year, or the number of arguments between spouses in one month. If the length of the time interval is the same for every individual in the sample, the methods already described work just fine. But if events are counted over different lengths of time for different individuals, there is clearly a need for some kind of standardization. For ordinary regression analysis, we could simply divide each individual's event count by the length of the interval and regress the resulting ratio on the independent variables. However, that won't work for Poisson regression because the division by time implies that the resulting variable no longer has a Poisson distribution.

Instead, we incorporate time into the model. If t_i is the length of the observation interval for individual *i*, the number of events (y_i) that occur during that interval is assumed to have the Poisson distribution

$$\Pr(y_i = r) = \frac{(\lambda_i t_i)^r e^{-\lambda_i t_i}}{r!}, \qquad r = 0, 1, 2, \ldots \tag{9.4}$$

which implies that the expected value of y_i is $\lambda_i t_i$. We continue to assume equation (9.3), which says that the logarithm of λ is a linear function of the x's. That implies that

$$\begin{aligned} \log E(y_i) &= \log(\lambda_i t_i) \\ &= \log(t_i) + \log(\lambda_i) \\ &= \log(t_i) + \beta_0 + \beta_1 x_{i1} + \beta_2 x_{i2} + \ldots + \beta_k x_{ik} \,. \end{aligned} \tag{9.5}$$

This equation says that the logarithm of the observation time should be on the right-hand side of the equation, with a coefficient of 1.0. Notice that if t_i is the same for everyone, this term can be absorbed into the intercept β_0.

Now it's easy enough to include log t as an independent variable in the regression model, but how do you force its coefficient to be 1.0? In GENMOD, you simply declare log t to be an OFFSET variable by using an option in the MODEL statement. Here's an example. Levinson et al. (1997) collected data on patient visits to 125 physicians. We'll take as our dependent variable the "number of utterances devoted by doctor or patient to prognostic material," which was obtained by investigator coding of 121 audio tapes (Christakis and Levinson 1998). The frequency distribution of this variable, called LENGTHPX, is shown in Output 9.10.

Output 9.10 Distribution of Number of Utterances about Prognosis

LENGTHPX	Frequency	Percent	Cumulative Frequency	Cumulative Percent
0	52	41.60	52	41.60
1	11	8.80	63	50.40
2	16	12.80	79	63.20
3	5	4.00	84	67.20
4	8	6.40	92	73.60
5	3	2.40	95	76.00
6	6	4.80	101	80.80
7	4	3.20	105	84.00
8	3	2.40	108	86.40
9	5	4.00	113	90.40
10	1	0.80	114	91.20
11	1	0.80	115	92.00
12	2	1.60	117	93.60
14	1	0.80	118	94.40
15	1	0.80	119	95.20
17	1	0.80	120	96.00
20	2	1.60	122	97.60
21	1	0.80	123	98.40
24	2	1.60	125	100.00

We'll consider the following independent variables:

PTAGE	Patient's age
PTSEX	Patient's sex (1=male, 0=female)
EZCOMPT	Doctor's rating of how easy it was to communicate with the patient (1-5)
MDLIKEPT	Doctor's rating of how much he or she liked the patient (1-5)
SURGEON	1 if doctor was a surgeon, otherwise 0
CLAIMS	Number of malpractice claims filed against the doctor
MINUTES	Length of the visit in minutes

To standardize for length of the session, we must define a new variable LMIN equal to the natural logarithm of MINUTES:

```
DATA prog2;
  SET prognosi;
  lmin=LOG(minutes);
RUN;
```

We then use LMIN as an OFFSET variable in the call to GENMOD:

```
PROC GENMOD DATA=prog2;
  MODEL lengthpx=ptage ptsex ezcompt mdlikept surgeon claims /
     OFFSET=lmin D=P;
RUN;
```

Results in Output 9.11 indicate that there are more utterances about prognosis when the patient is male, when the physician rates the patient easy to communicate with, when the physician is a surgeon, and when many malpractice claims have been filed against the physician. More specifically, surgeons have nearly four times as many prognosis utterances as non-surgeons (exp(1.34)=3.83) and there are about 73% more utterances for male patients than for females (exp(.55)=1.73). Notice that no coefficient is reported for LMIN because it is constrained to be 1.0.

Output 9.11 Poisson Regression with Offset Variable

Criteria For Assessing Goodness Of Fit			
Criterion	**DF**	**Value**	**Value/DF**
Deviance	114	682.0299	5.9827
Scaled Deviance	114	682.0299	5.9827
Pearson Chi-Square	114	899.6258	7.8915
Scaled Pearson X2	114	899.6258	7.8915
Log Likelihood		115.9561	
Full Log Likelihood		-455.9056	
AIC (smaller is better)		925.8112	
AICC (smaller is better)		926.8024	
BIC (smaller is better)		945.3818	

Analysis Of Maximum Likelihood Parameter Estimates							
Parameter	DF	Estimate	Standard Error	Wald 95% Confidence Limits		Wald Chi-Square	Pr > ChiSq
Intercept	1	-3.7237	0.3378	-4.3858	-3.0617	121.53	<.0001
ptage	1	-0.0014	0.0031	-0.0074	0.0046	0.22	0.6373
ptsex	1	0.5482	0.1048	0.3428	0.7537	27.35	<.0001
ezcompt	1	0.1981	0.0760	0.0490	0.3471	6.79	0.0092
mdlikept	1	-0.0864	0.0744	-0.2322	0.0593	1.35	0.2452
surgeon	1	1.3431	0.1304	1.0876	1.5986	106.14	<.0001
claims	1	0.0519	0.0232	0.0065	0.0974	5.01	0.0252
Scale	0	1.0000	0.0000	1.0000	1.0000		

Unfortunately, these results are called into question by the fact that the deviance is nearly six times the degrees of freedom, suggesting substantial overdispersion and consequent underestimation of the standard errors. To correct this problem, I reran the model with the negative binomial specification (D=NB), producing the results in Output 9.12.

Output 9.12 Negative Binomial Regression with Offset Variable

Criteria For Assessing Goodness Of Fit			
Criterion	DF	Value	Value/DF
Deviance	114	122.8819	1.0779
Scaled Deviance	114	122.8819	1.0779
Pearson Chi-Square	114	111.4206	0.9774
Scaled Pearson X2	114	111.4206	0.9774
Log Likelihood		304.9582	
Full Log Likelihood		-266.9035	
AIC (smaller is better)		549.8070	
AICC (smaller is better)		551.0927	
BIC (smaller is better)		572.1733	

Analysis Of Maximum Likelihood Parameter Estimates							
Parameter	DF	Estimate	Standard Error	Wald 95% Confidence Limits		Wald Chi-Square	Pr > ChiSq
Intercept	1	-3.3466	1.1372	-5.5755	-1.1177	8.66	0.0033
ptage	1	0.0011	0.0108	-0.0200	0.0223	0.01	0.9170
ptsex	1	0.5757	0.3013	-0.0148	1.1661	3.65	0.0560
ezcompt	1	0.1243	0.1954	-0.2588	0.5073	0.40	0.5249

Analysis Of Maximum Likelihood Parameter Estimates							
Parameter	DF	Estimate	Standard Error	Wald 95% Confidence Limits		Wald Chi-Square	Pr > ChiSq
mdlikept	1	-0.1036	0.2121	-0.5192	0.3120	0.24	0.6252
surgeon	1	1.4173	0.3681	0.6959	2.1388	14.83	0.0001
claims	1	0.0500	0.0768	-0.1006	0.2006	0.42	0.5154
Dispersion	1	2.1659	0.3866	1.5266	3.0730		

Note: The negative binomial dispersion parameter was estimated by maximum likelihood.

The negative binomial model fits much better, with a deviance only slightly greater than the degrees of freedom. In this model, only SURGEON is statistically significant at the .05 level, although PTSEX comes close.

9.7 Zero-Inflated Models

For some applications, the number of individuals with a count of zero may be a large fraction of the sample. For example, in the data set just examined, fully 42 percent of the patient visits had no "prognosis utterances." And for the biochemists that we looked at earlier, 60 percent published no articles when they were in graduate school. Is this a problem? Well, Poisson regression models often fit poorly when the fraction of zeros is large. This has led to the development of zero-inflated Poisson regression models which give special treatment to the zero counts. The zero-inflated Poisson (ZIP) model is now available in PROC GENMOD and PROC COUNTREG, along with a zero-inflated negative binomial model.

I must confess that I am not a huge fan of these models. In my experience, a conventional negative binomial model almost always fits better than a zero-inflated Poisson model. And since the ZIP model can be a lot harder to interpret, I would prefer to go with the negative binomial model. The zero-inflated negative binomial (ZINB) model will always fit at least a little better than the conventional negative binomial model (as assessed by the log-likelihood) because the latter is a special case of the former. But often the differences in fit between a conventional negative binomial and a zero-inflated negative binomial are quite small. So, again, I would prefer to avoid the conceptual complexities of the zero-inflated model.

There are some circumstances where a zero-inflated model makes a lot of sense, however, so it's worth taking the time to examine the model and to see how it's implemented and interpreted. Either of the two zero-inflated models starts with the assumption that the population consists of two groups:

1. For the *regression group*, a conventional Poisson or negative binomial regression model applies. Individuals in this group may or may not have count values of 0.
2. For the *zero group*, it's impossible to have a count other than 0. For these individuals, the covariates have no effect on the outcome.

An individual whose count is greater than zero is necessarily in the regression group. But an individual with an observed count of zero could be in either of the two groups.

This sort of model is sometimes called a finite mixture model. It can be estimated by maximum likelihood, even though we can't distinguish with certainty whether individuals with counts of zero are in one group or the other. In addition to the usual regression coefficients (for the individuals in the regression group), we can get an estimate of the probability that an individual is in the zero group. And we can elaborate the models further by allowing the probability of being in the zero group to be a function of covariates, usually via logistic regression.

Let's try these models on the patient visit data that we analyzed in the last section. Here's the GENMOD code for the basic zero-inflated Poisson model:

```
PROC GENMOD DATA=prog2;
  MODEL lengthpx=ptage ptsex ezcompt mdlikept surgeon claims /
     OFFSET=lmin D=ZIP;
  ZEROMODEL;
RUN;
```

As in the previous section, the OFFSET option is used to adjust for the varying lengths of the patient visits. To get the ZIP model, two changes are needed: D=ZIP on the MODEL statement, and a ZEROMODEL statement. As we'll see later, this statement allows one to specify covariates for the logistic regression part of the model. But it's required even when you don't want covariates.

Results are shown in Output 9.13. Although this model fits better than the conventional Poisson model in Output 9.11, the fit is still pretty bad, with a Pearson chi-square more than three times its degrees of freedom. Notice the new table at the end titled "Analysis of Maximum Likelihood Zero Inflation Parameter Estimates." For a model with no covariates in the ZEROMODEL statement, the value reported in this table (-.4839) is the estimated log-odds of being in the zero group. We can convert the log-odds to a probability by applying the logistic transformation, $1/(1+\exp(-\beta))$ which yields .38. So, according to this model, 38 percent of the sample members are estimated to be in the zero group, which has absolutely no chance of experiencing an event. Since 42 percent of the sample had a count of 0, that means that only 4 percent of the sample had counts of 0 but were not in the zero group. We can get a 95 percent confidence interval around the proportion by applying the logistic transformation to the upper and lower confidence limits in Output 9.13. This yields the interval (.29, .48).

Output 9.13 Zero-Inflated Poisson Model

Criteria For Assessing Goodness Of Fit			
Criterion	DF	Value	Value/DF
Deviance		691.2047	
Scaled Deviance		691.2047	
Pearson Chi-Square	113	350.9456	3.1057
Scaled Pearson X2	113	350.9456	3.1057
Log Likelihood		226.2593	
Full Log Likelihood		-345.6024	
AIC (smaller is better)		707.2047	
AICC (smaller is better)		708.4904	
BIC (smaller is better)		729.5710	

Analysis Of Maximum Likelihood Parameter Estimates							
Parameter	DF	Estimate	Standard Error	Wald 95% Confidence Limits		Wald Chi-Square	Pr > ChiSq
Intercept	1	-2.4835	0.3541	-3.1775	-1.7895	49.19	<.0001
ptage	1	-0.0039	0.0034	-0.0106	0.0028	1.28	0.2579
ptsex	1	0.3498	0.1119	0.1305	0.5691	9.77	0.0018
ezcompt	1	0.2639	0.0773	0.1124	0.4154	11.65	0.0006
mdlikept	1	-0.1527	0.0795	-0.3085	0.0031	3.69	0.0547
surgeon	1	0.8957	0.1445	0.6126	1.1789	38.45	<.0001
claims	1	0.0372	0.0228	-0.0075	0.0820	2.66	0.1029
Scale	0	1.0000	0.0000	1.0000	1.0000		

Analysis Of Maximum Likelihood Zero Inflation Parameter Estimates							
Parameter	DF	Estimate	Standard Error	Wald 95% Confidence Limits		Wald Chi-Square	Pr > ChiSq
Intercept	1	-0.4839	0.2034	-0.8827	-0.0852	5.66	0.0174

Table 9.1 displays full log-likelihoods and BIC statistics for this and several other models. We see that the ZIP model fits substantially better than the conventional Poisson, but not nearly as well as the conventional negative binomial (whose results we saw in Output 9.12).

Table 9.1 Measures of Fit for Several Models for Patient Visit Data

	Poisson	ZIP	NB	ZINB	ZIP w/ covariates	ZINB w/ covariates
Log-likelihood	-455.9	-345.6	-266.9	-265.8	-339.2	-261.0
BIC	945	730	572	575	746	594

We can fit the zero-inflated negative binomial model (without covariates) by changing the distribution option to read D=ZINB, which produces Output 9.14. As seen in Table 9.1, both the log-likelihood and BIC statistics for the ZINB model are much better than those for the ZIP model. The log-likelihood for the ZINB model is also slightly better than that for the conventional negative binomial, as it must be because the latter is a special case of the former. We can test the significance of the difference by computing a likelihood ratio chi-square (2 × the positive difference in the log-likelihoods) which is only 2.2 with 1 d.f., far from statistically significant. This result is corroborated by the fact that the BIC statistic for the conventional negative binomial is better than the BIC for the ZINB model. That's because BIC penalizes the ZINB model for the extra parameter in the model.

The coefficients and *p*-values in the first table of Output 9.14 are very similar to those for the conventional negative binomial in Output 9.12. The dispersion parameter is noticeably lower, however, probably because the zero parameter allows for some of the overdispersion. The zero parameter estimate is also considerably lower than the estimate for the ZIP model in Output 9.13. Applying the logistic transformation to -1.2757, we get an estimated probability of .22 for being in the zero group.

Output 9.14 Zero-Inflated Negative Binomial Model

Analysis Of Maximum Likelihood Parameter Estimates							
Parameter	DF	Estimate	Standard Error	Wald 95% Confidence Limits		Wald Chi-Square	Pr > ChiSq
Intercept	1	-2.7212	1.0854	-4.8486	-0.5938	6.29	0.0122
ptage	1	-0.0028	0.0104	-0.0230	0.0175	0.07	0.7904
ptsex	1	0.5382	0.2730	0.0032	1.0731	3.89	0.0487
ezcompt	1	0.1445	0.1743	-0.1970	0.4860	0.69	0.4070
mdlikept	1	-0.1276	0.1925	-0.5048	0.2496	0.44	0.5072
surgeon	1	1.2349	0.3680	0.5137	1.9561	11.26	0.0008
claims	1	0.0441	0.0669	-0.0869	0.1751	0.43	0.5096
Dispersion	1	1.1759	0.4441	0.5609	2.4652		

Analysis Of Maximum Likelihood Zero Inflation Parameter Estimates							
Parameter	DF	Estimate	Standard Error	Wald 95% Confidence Limits		Wald Chi-Square	Pr > ChiSq
Intercept	1	-1.2757	0.5883	-2.4288	-0.1227	4.70	0.0301

As noted, you can also fit a ZIP or ZINB model with covariates that predict the probability of being in the zero group. This is accomplished simply by listing the desired covariates on the ZEROMODEL statement. The covariates can be the same as those for the main regression model or they can be different—there is no restriction. Here is the code for fitting a ZINB model using the same covariates on the MODEL and ZEROMODEL statements:

```
PROC GENMOD DATA=prog2;
  MODEL lengthpx=ptage ptsex ezcompt mdlikept surgeon claims /
     OFFSET=lmin D=ZINB;
  ZEROMODEL ptage ptsex ezcompt mdlikept surgeon claims;
RUN;
```

Output 9.15 gives the results. The only covariate in the zero model with a *p*-value below .05 is SURGEON, which is also the only significant variable in the main model. But the *p*-value in the main model is much higher than it was in the conventional negative binomial model. It appears that the effect of SURGEON has been distributed across the two portions of the model. Being a surgeon both increases the expected number of prognostic utterances (among those not in the zero group), and also reduces the probability of being in the zero group. As shown in Table 9.1, this model has the highest log-likelihood (as expected). But its BIC statistic is worse than both the conventional negative binomial model and the ZINB model without covariates. I also fit a ZIP model with covariates (detailed results not shown), but it clearly fit much worse than any of the negative binomial models.

Output 9.15 Zero-Inflated Negative Binomial Model with Covariates

Analysis Of Maximum Likelihood Parameter Estimates							
Parameter	DF	Estimate	Standard Error	Wald 95% Confidence Limits		Wald Chi-Square	Pr > ChiSq
Intercept	1	-1.5862	1.0070	-3.5599	0.3875	2.48	0.1152
ptage	1	-0.0143	0.0111	-0.0361	0.0075	1.65	0.1995
ptsex	1	0.4939	0.2872	-0.0691	1.0568	2.96	0.0855
ezcompt	1	0.1588	0.1758	-0.1858	0.5035	0.82	0.3664
mdlikept	1	-0.1699	0.1980	-0.5579	0.2182	0.74	0.3910
surgeon	1	0.7669	0.3579	0.0654	1.4685	4.59	0.0321
claims	1	0.0530	0.0656	-0.0756	0.1817	0.65	0.4190
Dispersion	1	1.0321	0.3407	0.5404	1.9712		

Analysis Of Maximum Likelihood Zero Inflation Parameter Estimates							
Parameter	DF	Estimate	Standard Error	Wald 95% Confidence Limits		Wald Chi-Square	Pr > ChiSq
Intercept	1	3.3612	2.4623	-1.4648	8.1872	1.86	0.1722
ptage	1	-0.0454	0.0326	-0.1093	0.0185	1.94	0.1638
ptsex	1	-0.6221	0.6590	-1.9138	0.6696	0.89	0.3452
ezcompt	1	0.1910	0.4443	-0.6798	1.0617	0.18	0.6673
mdlikept	1	-0.2443	0.4674	-1.1604	0.6717	0.27	0.6011
surgeon	1	-2.0087	0.8906	-3.7543	-0.2632	5.09	0.0241
claims	1	-0.0115	0.1613	-0.3276	0.3046	0.01	0.9431

So, for these data, the winner is the conventional negative binomial model. It fits the data well and has a lower BIC statistic than any of the zero-inflated models. It’s also conceptually simpler than any of the inflated models.

To sum up, the conventional negative binomial model is almost always better than the zero-inflated Poisson model. The zero-inflated negative binomial model may sometimes fit better than the conventional negative binomial, but for many applications it does not. It's important to test for the significance of the difference.

As previously mentioned, however, there are some applications in which a compelling case can be made for a zero-inflated model. Suppose, for example, that the sample consists of 50-year-old women, and the variable of interest is number of children ever born. It is reasonable to believe that at least some of those women were biologically sterile. These women necessarily have value of 0 on the dependent variable, and any variation in their covariate values cannot affect the outcome. In this case, a zero-inflated model would make a lot of sense. But if you're going that route, I would strongly recommend a ZINB model rather than a ZIP model. And I would still want to see evidence that the ZINB model fits better than the conventional negative binomial model.

Chapter 10

Loglinear Analysis of Contingency Tables

10.1 Introduction

In Chapter 4, we saw how to analyze contingency tables that included a dichotomous variable treated as dependent on the other variables. The strategy was to directly estimate a logit model in PROC LOGISTIC. We extended that approach in Chapter 5 to handle dependent variables with more than two categories by estimating a multinomial logit model with PROC LOGISTIC. In this chapter, we see how to estimate *loglinear* models for contingency tables. What distinguishes loglinear models from logit models is that loglinear models do not have an explicit dependent variable, at least not one that corresponds to any conceptual variable. As I've mentioned previously, every logit model for a contingency table has a loglinear model that is exactly equivalent. But the class of loglinear models also includes models that do *not* correspond to any logit models, so we are dealing with a much wider class of models.

Why do we need loglinear models? When loglinear analysis was first developed in the late 1960s and early 1970s, there wasn't much software available for logistic regression. And what *was* available wasn't very suitable for analyzing contingency tables. By contrast, loglinear models were easily estimated with widely available software. Nowadays, however, you're usually better off estimating a logit model directly. Logit models are simpler and correspond more directly to substantive theory. However, there are some situations when a logit model just won't do the job. For example, you might want to explore the relationships among several attitude measures that have no obvious causal ordering. But a logit model requires that you choose one variable as the dependent variable and the others as independent variables.

Even if you do have a clear-cut dependent variable, the particular logit model you want to estimate may be awkward for conventional software. For example, the adjacent-categories model described in Chapter 6 is a special case of the multinomial logit model. But the maximum likelihood algorithm used in PROC LOGISTIC will not impose the necessary constraints to get this model. As we shall see, the adjacent-categories model is easily estimated as a loglinear model when the data come in the form of a contingency table. Loglinear models are particularly well suited to the analysis of two-way tables in which the row variable has the same categories as the column variables (a square table). For example, there is a large literature on loglinear models for mobility tables in which the row variable represents parent's occupation and the column variable represents child's occupation (for example, Hout 1983).

The treatment of loglinear analysis in this book is far from comprehensive. The topic is so vast that I can do no more than scratch the surface here. My goals are, first, to give you some idea of what loglinear models are like and how they are related to logit models. Second, I will show you how to estimate some fairly conventional loglinear models by using the GENMOD procedure. Third, I will present some examples in which a loglinear model has significant advantages over a logit model.

10.2 A Loglinear Model for a 2 × 2 Table

Let's start with the 2 × 2 table we analyzed in Chapter 4 (reproduced here as Table 10.1), which shows sentence by race for 147 death penalty cases. How can we represent this by a loglinear model?

Table 10.1 Death Sentence by Race of Defendant

	Blacks	Nonblacks	Total
Death	28	22	50
Life	45	52	97
Total	73	74	147

Let's consider the table in more general form as

m_{11}	m_{12}
m_{21}	m_{22}

where m_{ij} is the *expected* number of cases falling into row *i* and column *j*. What I mean by this is that if *n* is the sample size and p_{ij} is the probability of falling into cell (*i*, *j*), then $m_{ij}=np_{ij}$.

There are a couple of different but equivalent ways of writing a loglinear model for these four frequency counts. The way I'm going to do it is consistent with how PROC GENMOD estimates the model. Let R_i be a dummy variable for the rows, having a value of 1 if i=1 and 0 if i=2. Similarly, let C_j be a dummy variable for the columns with a value of 1 if j=1 and 0 if j=2. We can then write the "saturated" loglinear model for this 2×2 table of frequency counts as

$$\log m_{ij} = \beta_0 + \beta_1 R_i + \beta_2 C_j + \beta_3 R_i C_j \qquad i, j = 1, 2. \tag{10.1}$$

Note that R_iC_j is the product (interaction) of R_i and C_j. Equation (10.1) actually represents four different equations:

$$\begin{aligned} \log m_{11} &= \beta_0 + \beta_1 + \beta_2 + \beta_3 \\ \log m_{12} &= \beta_0 + \beta_1 \\ \log m_{21} &= \beta_0 \qquad + \beta_2 \\ \log m_{22} &= \beta_0 . \end{aligned} \tag{10.2}$$

In a moment, we'll see how to estimate the four β parameters using PROC GENMOD. But before doing that, let's talk about why we might be interested in these parameters at all. All we've done in equation (10.2) is transform the four expected frequencies into four different quantities. The reason for doing this is that the β's show us things about the table that aren't so easily seen by looking at the frequency counts.

We are particularly interested in β_3, which is the coefficient for the interaction term. If we solve equations (10.2) for β_3, we get:

$$\beta_3 = \log\left(\frac{m_{11}m_{22}}{m_{12}m_{21}}\right) \quad (10.3)$$

The quantity in parenthesis is the cross-product ratio. In Chapter 2, we saw that the cross-product ratio is equal to the odds ratio. In this case, it's the ratio of the odds of a death sentence for blacks to the odds of a death sentence for nonblacks. Recall that an odds ratio of 1.0 corresponds to independence between the two variables. Because the logarithm of 1 is 0, independence of the row and column variables is equivalent to β_3=0. So, we can test whether the two variables are independent by testing whether β_3=0.

All this has been expressed in terms of *expected* frequency counts, the m_{ij}. What about the observed frequency counts, which I'll denote by n_{ij}? For the model we've just been considering, the maximum likelihood estimator of β_3 has the same form as equation (10.3) except that expected frequencies are replaced by observed frequencies:

$$\hat{\beta}_3 = \log\left(\frac{n_{11}n_{22}}{n_{12}n_{21}}\right) \quad (10.4)$$

For the data in Table 10.1, $\hat{\beta}_3 = \log[(28\times 52)/(45\times 22)] = .3857$. Similar expressions can readily be obtained for the other β parameters in the model, but we're usually not very interested in them.

Although it's simple enough in this case to get all the maximum likelihood estimates by hand calculations, we can also do it with PROC GENMOD. Here is the SAS code:

```
DATA penalty;
  INPUT n death black;
  DATALINES;
28 1 1
22 1 0
45 0 1
52 0 0
;
PROC GENMOD DATA=penalty;
  MODEL n = death black death*black / DIST=POISSON;
RUN;
```

Each cell in the table is a separate record in the data set. DEATH is a dummy variable for a death sentence, and BLACK is a dummy variable for race. The MODEL statement includes the main effects of row and column, as well as their interaction. The DIST option in the MODEL statement says that each frequency count has a Poisson distribution whose expected value m_{ij} is given by equation (10.2). For frequency counts in contingency tables, the Poisson distribution is appropriate for a variety of different sampling designs. As we saw in Chapter 9, the default in GENMOD for a Poisson distribution is LINK=LOG. That means that the logarithm of the expected value of the dependent variable is assumed to be a linear function of the explanatory variables, which is exactly what equation (10.1) says.

Output 10.1 PROC GENMOD Output for Loglinear Model for a 2 × 2 Table

Criteria For Assessing Goodness Of Fit			
Criterion	**DF**	**Value**	**Value/DF**
Deviance	0	0.0000	.
Scaled Deviance	0	0.0000	.
Pearson Chi-Square	.	0.0000	.
Scaled Pearson X2	.	0.0000	.
Log Likelihood		391.0691	
Full Log Likelihood		-10.7765	
AIC (smaller is better)		29.5531	
AICC (smaller is better)		.	
BIC (smaller is better)		27.0983	

Analysis Of Maximum Likelihood Parameter Estimates							
Parameter	**DF**	**Estimate**	**Standard Error**	**Wald 95% Confidence Limits**		**Wald Chi-Square**	**Pr > ChiSq**
Intercept	1	3.9512	0.1387	3.6794	4.2230	811.84	<.0001
death	1	-0.8602	0.2543	-1.3587	-0.3617	11.44	0.0007
black	1	-0.1446	0.2036	-0.5436	0.2545	0.50	0.4776
death*black	1	0.3857	0.3502	-0.3006	1.0721	1.21	0.2706
Scale	0	1.0000	0.0000	1.0000	1.0000		

Examining Output 10.1, we see that the deviance for this model is 0. That's because it's a *saturated* model; there are four estimated parameters for the four cells in the table, so the model perfectly reproduces the frequency counts. The estimate for the interaction is .3857, which is the same value obtained from hand calculation of equation (10.4). This estimate has an associated Wald chi-square of 1.214, which is nearly identical to the traditional Pearson chi-square (1.218) for testing whether the two variables are independent (obtained with PROC FREQ). That's not surprising because both statistics are testing the same null hypothesis.

Instead of a loglinear model, we could estimate a logit model for this table, taking death sentence as the dependent variable:

```
PROC GENMOD DATA=penalty DESC;
  FREQ n;
  MODEL death=black / D=BINOMIAL LINK=LOGIT AGGREGATE;
RUN;
```

This produces Output 10.2.

Output 10.2 PROC GENMOD Output for Logit Model for a 2 × 2 Table

Analysis Of Maximum Likelihood Parameter Estimates							
Parameter	DF	Estimate	Standard Error	Wald 95% Confidence Limits		Wald Chi-Square	Pr > ChiSq
Intercept	1	-0.8602	0.2543	-1.3587	-0.3617	11.44	0.0007
black	1	0.3857	0.3502	-0.3006	1.0721	1.21	0.2706
Scale	0	1.0000	0.0000	1.0000	1.0000		

Remarkably, the estimates in Output 10.2 are identical to some of the estimates in Output 10.1. The coefficient for BLACK in the logit model is the same as the BLACK*DEATH coefficient in the loglinear model, along with identical standard errors and chi-squares. Similarly, the intercept (and associated statistics) in the logit model is the same as the DEATH coefficient in the loglinear model. This is a general phenomenon. As I've said before, every logit model for a contingency table has a corresponding loglinear model. But the main effects in the logit model become 2-way interactions (with the dependent variable) in the loglinear model. The intercept in the logit model becomes a main effect of the dependent variable in the loglinear model.

Now let's do something peculiar just to emphasize the point. We'll estimate another logit model, but we'll switch the variables around and make BLACK the dependent variable and DEATH the independent variable:

```
PROC GENMOD DATA=penalty DESC;
  FREQ n;
  MODEL black=death / D=BINOMIAL LINK=LOGIT AGGREGATE;
RUN;
```

Output 10.3 PROC GENMOD Output for Logit Model for a 2 × 2 Table with Variables Reversed

Analysis Of Maximum Likelihood Parameter Estimates							
Parameter	DF	Estimate	Standard Error	Wald 95% Confidence Limits		Wald Chi-Square	Pr > ChiSq
Intercept	1	-0.1446	0.2036	-0.5436	0.2545	0.50	0.4776
death	1	0.3857	0.3502	-0.3006	1.0721	1.21	0.2706
Scale	0	1.0000	0.0000	1.0000	1.0000		

In Output 10.3, we see again that the slope coefficient (along with its standard error and chi-square) is identical to the two-way interaction in the loglinear model. Unlike bivariate linear regression, where the regression line changes when you reverse the dependent and independent variables, the bivariate logit model is symmetrical with respect to the two variables. We also see that the intercept term corresponds to the main effect of BLACK in the loglinear model.

Why do the logit models share parameters with the loglinear model? The algebra that demonstrates this is really quite simple. The dependent variable in the logit model is the logarithm of the odds. As explained in chapter 2, the odds can be defined as the expected number of events divided by the expected number of non-events. For example, the log-odds of a death sentence for racial group j is $\log(m_{1j}/m_{2j})$. Substituting from the loglinear model in equation (10.1), we get:

$$\log\left(\frac{m_{1j}}{m_{2j}}\right) = \log(m_{1j}) - \log(m_{2j})$$
$$= \left(\beta_0 + \beta_1 R_1 + \beta_2 C_j + \beta_3 R_1 C_j\right) - \left(\beta_0 + \beta_1 R_2 + \beta_2 C_j + \beta_3 R_2 C_j\right).$$

But because R_1=1 and R_2=0, this reduces to:

$$\log\left(\frac{m_{1j}}{m_{2j}}\right) = \beta_1 + \beta_3 C_j.$$

Similarly, if we reverse the independent and dependent variables, we obtain:

$$\log\left(\frac{m_{i1}}{m_{i2}}\right) = \beta_2 + \beta_3 R_i.$$

These results show that the loglinear model for the 2×2 table implies two logit models, one for the row variable as dependent on the column variable and the other for the column variable as dependent on the row variable.

10.3 Loglinear Models for a Four-Way Table

Now let's look at a much more complicated table, the $2 \times 2 \times 4 \times 4$ table that we previously analyzed by way of a logit model in Section 4.5. Our main goal will be to duplicate the results of the logit model with a loglinear model. To refresh your memory, the sample consisted of 4,991 high school seniors in Wisconsin. The dependent variable was whether or not they planned to attend college in the following year. The three independent variables were coded as follows:

IQ	1=low, 2=lower middle, 3=upper middle, 4=high
SES	1=low, 2=lower middle, 3=upper middle, 4=high
PARENT	1=low parental encouragement, 2= high encouragement

In Section 4.5, the data were read in as 32 records, each record containing a unique combination of values of the independent variables, along with the number of seniors who had those values and the number of those seniors who planned to attend college. Unfortunately, that's not the format we need for a loglinear analysis. Instead, we need 64 records, one for each cell in the four-way table, with values for all the variables and the frequency count in that cell. Here's a DATA step that inputs the previous data set (WISC) and outputs the new data set in the appropriate format (WISCTAB).

```
DATA wisctab;
  SET wisc;
  college=1;
  freq=coll;
  OUTPUT;
  college=0;
  freq=total-coll;
  OUTPUT;
  DROP total coll;
PROC PRINT;
RUN;
```

Output 10.4 shows what this new data set looks like.

Output 10.4 Data for a Four-Way Contingency Table

Obs	iq	parent	ses	college	freq
1	1	1	1	1	4
2	1	1	1	0	349
3	1	1	2	1	2
4	1	1	2	0	232
5	1	1	3	1	8
6	1	1	3	0	166
7	1	1	4	1	4
8	1	1	4	0	48
9	1	2	1	1	13
10	1	2	1	0	64
11	1	2	2	1	27
12	1	2	2	0	84
13	1	2	3	1	47
14	1	2	3	0	91
15	1	2	4	1	39
16	1	2	4	0	57
17	2	1	1	1	9
18	2	1	1	0	207
19	2	1	2	1	7
20	2	1	2	0	201
21	2	1	3	1	6
22	2	1	3	0	120
23	2	1	4	1	5
24	2	1	4	0	47
25	2	2	1	1	33
26	2	2	1	0	72
27	2	2	2	1	64
28	2	2	2	0	95
29	2	2	3	1	74
30	2	2	3	0	110
31	2	2	4	1	123
32	2	2	4	0	90
33	3	1	1	1	12
34	3	1	1	0	126
35	3	1	2	1	12

Obs	iq	parent	ses	college	freq
36	3	1	2	0	115
37	3	1	3	1	17
38	3	1	3	0	92
39	3	1	4	1	9
40	3	1	4	0	41
41	3	2	1	1	38
42	3	2	1	0	54
43	3	2	2	1	93
44	3	2	2	0	92
45	3	2	3	1	148
46	3	2	3	0	100
47	3	2	4	1	224
48	3	2	4	0	65
49	4	1	1	1	10
50	4	1	1	0	67
51	4	1	2	1	17
52	4	1	2	0	79
53	4	1	3	1	6
54	4	1	3	0	42
55	4	1	4	1	8
56	4	1	4	0	17
57	4	2	1	1	49
58	4	2	1	0	43
59	4	2	2	1	119
60	4	2	2	0	59
61	4	2	3	1	198
62	4	2	3	0	73
63	4	2	4	1	414
64	4	2	4	0	54

Here is the SAS code for estimating a loglinear model that is equivalent to the first logit model of Section 4.5:

```
PROC GENMOD DATA=wisctab;
  CLASS iq ses;
  MODEL freq=iq|ses|parent college iq*college ses*college
       parent*college / D=P TYPE3;
RUN;
```

As before, we are fitting a Poisson regression model for the frequency counts, with the default logarithmic link. The first term on the right-hand side of the MODEL equation—IQ|SES|PARENT—is shorthand for IQ*SES*PARENT IQ*SES IQ*PARENT SES*PARENT IQ SES PARENT. In other words, we fit the 3-way interaction, the three 2-way interactions, and the main effects of each of the independent variables. These parameters pertain only to the relationships among the independent variables in the logit model, not to the effects of the independent variables on the dependent variable (college choice). We include them in the model because to do otherwise would assert that they are 0. Because we cannot force these parameters to be 0 in a logit model, neither do we do it in the corresponding loglinear model. The general principle is this: *Whenever you want a loglinear model to be equivalent to some logit model, you must include all possible interactions among the predictor variables in the logit model.* Even though we include these interactions, they rarely have any substantive interest because they describe relationships among the predictor variables *conditional on the values of the dependent variable*. Ordinarily, these parameters have no useful causal interpretation.

The parameters that do have a useful interpretation are specified in the MODEL statement as COLLEGE IQ*COLLEGE SES*COLLEGE PARENT*COLLEGE. These correspond to the intercept and the three main effects of the independent variables on the dependent variable in the logit model. So, all the parameters in the corresponding logit model involve the dependent variable when specified in the loglinear model. Notice that IQ and SES are listed as CLASS variables so that, for each variable, three dummy variables will be constructed to represent the four categories. This is unnecessary for COLLEGE and PARENT because they are dichotomous.

Results are shown in Output 10.5. Values that are the same as those in Output 4.10, obtained by direct fitting of the logit model, are shown in italics. The numbers to the right of the parameter names correspond to the values of the CLASS variables. Apparently, the loglinear model contains many more parameters than the logit model, but the parameters that matter are identical in the two models. Notice also that the deviance and Pearson chi-squares are identical for the logit and loglinear models. The Type 3 chi-squares are not identical in the two outputs because PROC LOGISTIC reports Wald chi-squares while PROC GENMOD reports likelihood ratio chi-squares. But the values are similar.

Output 10.5 PROC GENMOD Output for Loglinear Analysis of a Four-Way Table

Criteria For Assessing Goodness Of Fit			
Criterion	DF	Value	Value/DF
Deviance	24	25.2358	1.0515
Scaled Deviance	24	25.2358	1.0515
Pearson Chi-Square	24	24.4398	1.0183
Scaled Pearson X2	24	24.4398	1.0183
Log Likelihood		18912.8805	
Full Log Likelihood		-192.6604	
AIC (smaller is better)		465.3207	
AICC (smaller is better)		607.9294	
BIC (smaller is better)		551.6760	

Analysis Of Maximum Likelihood Parameter Estimates									
Parameter			DF	Estimate	Standard Error	Wald 95% Confidence Limits		Wald Chi-Square	Pr > ChiSq
Intercept			1	1.4076	0.4093	0.6055	2.2098	11.83	0.0006
iq	1		1	2.4070	0.5036	1.4200	3.3939	22.85	<.0001
iq	2		1	1.8198	0.4967	0.8462	2.7934	13.42	0.0002
iq	3		1	1.7044	0.4969	0.7305	2.6782	11.77	0.0006
iq	4		0	0.0000	0.0000	0.0000	0.0000	.	.
ses	1		1	3.4289	0.4775	2.4929	4.3648	51.55	<.0001
ses	2		1	3.3431	0.4600	2.4416	4.2446	52.83	<.0001
ses	3		1	1.6470	0.5007	0.6657	2.6283	10.82	0.0010
ses	4		0	0.0000	0.0000	0.0000	0.0000	.	.
iq*ses	1	1	1	0.2993	0.5817	-0.8408	1.4393	0.26	0.6069
iq*ses	1	2	1	-0.7469	0.5692	-1.8625	0.3687	1.72	0.1895
iq*ses	1	3	1	0.2001	0.6061	-0.9879	1.3881	0.11	0.7413
iq*ses	1	4	0	0.0000	0.0000	0.0000	0.0000	.	.
iq*ses	2	1	1	-0.3036	0.5794	-1.4392	0.8321	0.27	0.6004
iq*ses	2	2	1	-0.6241	0.5633	-1.7282	0.4799	1.23	0.2679
iq*ses	2	3	1	0.0101	0.6070	-1.1796	1.1999	0.00	0.9867
iq*ses	2	4	0	0.0000	0.0000	0.0000	0.0000	.	.
iq*ses	3	1	1	-0.7575	0.5905	-1.9148	0.3999	1.65	0.1996
iq*ses	3	2	1	-1.4171	0.5746	-2.5433	-0.2910	6.08	0.0137
iq*ses	3	3	1	-0.2082	0.6113	-1.4063	0.9899	0.12	0.7334
iq*ses	3	4	0	0.0000	0.0000	0.0000	0.0000	.	.
iq*ses	4	1	0	0.0000	0.0000	0.0000	0.0000	.	.

Analysis Of Maximum Likelihood Parameter Estimates									
Parameter			DF	Estimate	Standard Error	Wald 95% Confidence Limits		Wald Chi-Square	Pr > ChiSq
iq*ses	4	2	0	0.0000	0.0000	0.0000	0.0000	.	.
iq*ses	4	3	0	0.0000	0.0000	0.0000	0.0000	.	.
iq*ses	4	4	0	0.0000	0.0000	0.0000	0.0000	.	.
parent			1	1.3895	0.2144	0.9693	1.8097	42.01	<.0001
parent*iq	1		1	-1.3237	0.2746	-1.8619	-0.7855	23.24	<.0001
parent*iq	2		1	-0.7906	0.2632	-1.3064	-0.2747	9.02	0.0027
parent*iq	3		1	-0.8352	0.2616	-1.3479	-0.3225	10.19	0.0014
parent*iq	4		0	0.0000	0.0000	0.0000	0.0000	.	.
parent*ses	1		1	-2.0023	0.2641	-2.5200	-1.4846	57.46	<.0001
parent*ses	2		1	-1.7432	0.2474	-2.2281	-1.2583	49.64	<.0001
parent*ses	3		1	-0.7940	0.2636	-1.3107	-0.2774	9.07	0.0026
parent*ses	4		0	0.0000	0.0000	0.0000	0.0000	.	.
parent*iq*ses	1	1	1	0.2425	0.3363	-0.4167	0.9017	0.52	0.4709
parent*iq*ses	1	2	1	0.6967	0.3196	0.0704	1.3231	4.75	0.0292
parent*iq*ses	1	3	1	0.1915	0.3316	-0.4584	0.8415	0.33	0.5636
parent*iq*ses	1	4	0	0.0000	0.0000	0.0000	0.0000	.	.
parent*iq*ses	2	1	1	0.3940	0.3240	-0.2410	1.0291	1.48	0.2239
parent*iq*ses	2	2	1	0.4903	0.3061	-0.1096	1.0901	2.57	0.1092
parent*iq*ses	2	3	1	0.0860	0.3228	-0.5468	0.7187	0.07	0.7900
parent*iq*ses	2	4	0	0.0000	0.0000	0.0000	0.0000	.	.
parent*iq*ses	3	1	1	0.5263	0.3284	-0.1173	1.1698	2.57	0.1090
parent*iq*ses	3	2	1	0.9028	0.3082	0.2987	1.5069	8.58	0.0034
parent*iq*ses	3	3	1	0.2568	0.3215	-0.3733	0.8869	0.64	0.4244
parent*iq*ses	3	4	0	0.0000	0.0000	0.0000	0.0000	.	.
parent*iq*ses	4	1	0	0.0000	0.0000	0.0000	0.0000	.	.
parent*iq*ses	4	2	0	0.0000	0.0000	0.0000	0.0000	.	.
parent*iq*ses	4	3	0	0.0000	0.0000	0.0000	0.0000	.	.
parent*iq*ses	4	4	0	0.0000	0.0000	0.0000	0.0000	.	.
college			*1*	*-3.1005*	*0.2123*	*-3.5166*	*-2.6845*	*213.34*	*<.0001*
*college*iq*	*1*		*1*	*-1.9663*	*0.1210*	*-2.2034*	*-1.7293*	*264.24*	*<.0001*
*college*iq*	*2*		*1*	*-1.3722*	*0.1024*	*-1.5728*	*-1.1716*	*179.73*	*<.0001*
*college*iq*	*3*		*1*	*-0.6331*	*0.0976*	*-0.8244*	*-0.4418*	*42.08*	*<.0001*
*college*iq*	*4*		*0*	*0.0000*	*0.0000*	*0.0000*	*0.0000*	*.*	*.*
*college*ses*	*1*		*1*	*-1.4140*	*0.1210*	*-1.6510*	*-1.1769*	*136.67*	*<.0001*
*college*ses*	*2*		*1*	*-1.0580*	*0.1029*	*-1.2596*	*-0.8564*	*105.79*	*<.0001*
*college*ses*	*3*		*1*	*-0.7516*	*0.0976*	*-0.9428*	*-0.5604*	*59.34*	*<.0001*

Analysis Of Maximum Likelihood Parameter Estimates									
Parameter			DF	Estimate	Standard Error	Wald 95% Confidence Limits		Wald Chi-Square	Pr > ChiSq
*college*ses*	*4*		*0*	*0.0000*	*0.0000*	*0.0000*	*0.0000*	*.*	*.*
*parent*college*			*1*	*2.4554*	*0.1014*	*2.2567*	*2.6541*	*586.39*	*<.0001*
Scale			0	1.0000	0.0000	1.0000	1.0000		

LR Statistics For Type 3 Analysis			
Source	DF	Chi-Square	Pr > ChiSq
iq	3	175.60	<.0001
ses	3	379.72	<.0001
iq*ses	9	17.56	0.0406
parent	1	34.12	<.0001
parent*iq	3	86.16	<.0001
parent*ses	3	257.09	<.0001
parent*iq*ses	9	13.73	0.1321
college	1	1078.37	<.0001
college*iq	3	361.56	<.0001
college*ses	3	179.85	<.0001
parent*college	1	795.61	<.0001

Because the deviance is not 0, we know that this is not a saturated model, unlike the model we considered for the 2×2 table. To get a saturated model, we would have to include three 3-way interactions with COLLEGE and one 4-way interaction with COLLEGE. These would correspond to three 2-way interactions and one 3-way interaction in the logit model.

10.4 Fitting the Adjacent Categories Model as a Loglinear Model

Now we know how to fit a logit model to a contingency table by fitting an equivalent loglinear model, but a natural question is, "Why bother?" As we've just seen, the loglinear model is cluttered with "nuisance parameters" and is more cumbersome to specify in the MODEL statement. Even worse, the requirement of fitting the full multi-way interaction for the independent variables often leads to annoying convergence problems. Specifically, if there are any cell frequencies of 0 in the multi-way contingency table for the independent variables, the model will not converge. This problem does not arise when fitting the logit model directly.

Despite these difficulties, there are situations in which it is easy to fit the loglinear model but difficult to fit the corresponding logit model. One such model is the adjacent-categories model that we considered in Chapter 6. Recall that the adjacent-categories model is one of three alternative logit models for dependent variables with multiple, ordered categories. If the dependent variable has *J* ordered categories, the model says that for individual *i*:

$$\log\left(\frac{p_{ij}}{p_{i,j+1}}\right)=\alpha_j+\boldsymbol{\beta}\mathbf{x}_i \qquad j=1,\ldots,J-1$$

In other words, if we contrast adjacent categories of the dependent variable, the effect of the independent variables on their relative odds is the same regardless of which pair we consider.

This model is a special case of the multinomial logit model that does not impose any ordering on the categories of the dependent variable. (We get the general model by putting a *j* subscript on the $\boldsymbol{\beta}$ coefficient.) In Chapter 6, we saw how we could estimate the adjacent categories model in PROC CATMOD by the method of weighted least squares. But the maximum likelihood algorithm in CATMOD cannot impose the necessary constraints on the multinomial model.

Now, we'll see how to easily estimate the adjacent categories model as a loglinear model by maximum likelihood. In Chapter 6, we looked at the effect of calendar year (1974, 1984, 1994) and marital status (married, unmarried) on reported happiness (very happy=1, pretty happy=2, not too happy=3). Section 6.1 displays the contingency table and the SAS code for reading in the data. But we need to modify the data set as follows:

```
DATA happy2;
  SET happy;
  happyq=happy;
RUN;
```

In this DATA step, I defined a new variable HAPPYQ to be identical to HAPPY, the three-category dependent variable. By defining two versions of the dependent variable, we can treat it as both quantitative and qualitative in the same model. The model is specified as follows:

```
PROC GENMOD DATA=happy2;
  CLASS year happy;
  MODEL count=year|married happy year*happyq married*happyq /
    D=P;
RUN;
```

HAPPY and YEAR are declared as CLASS variables, but HAPPYQ is treated as a quantitative variable. Because MARRIED has only two categories, it doesn't need to be a CLASS variable. The term YEAR|MARRIED ensures that the model is equivalent to a logit model by fitting all possible terms pertaining to the two independent variables. The effects of YEAR and MARRIED on happiness are specified by the terms YEAR*HAPPYQ and MARRIED*HAPPYQ. By also including HAPPY in the model as a CLASS variable, we allow for different intercepts for the two adjacent pairs of categories. Results are shown in Output 10.6. As in Output 10.5, those parameters that correspond to the equivalent logit model are shown in italics.

Output 10.6 PROC GENMOD Output for the Adjacent Categories Model

Criteria For Assessing Goodness Of Fit			
Criterion	**DF**	**Value**	**Value/DF**
Deviance	7	35.7555	5.1079
Scaled Deviance	7	35.7555	5.1079
Pearson Chi-Square	7	35.9250	5.1321
Scaled Pearson X2	7	35.9250	5.1321
Log Likelihood		28747.4570	
Full Log Likelihood		-83.6319	
AIC (smaller is better)		189.2639	
AICC (smaller is better)		233.2639	
BIC (smaller is better)		199.0580	

Analysis Of Maximum Likelihood Parameter Estimates								
Parameter		**DF**	**Estimate**	**Standard Error**	**Wald 95% Confidence Limits**		**Wald Chi-Square**	**Pr > ChiSq**
Intercept		1	5.6206	0.0492	5.5241	5.7171	13029.6	<.0001
year	1	1	-1.1262	0.1182	-1.3580	-0.8945	90.71	<.0001
year	2	1	-0.6319	0.1144	-0.8562	-0.4075	30.48	<.0001
year	3	0	0.0000	0.0000	0.0000	0.0000	.	.
married		1	1.5386	0.0918	1.3587	1.7184	281.07	<.0001
married*year	1	1	0.8596	0.0710	0.7205	0.9986	146.78	<.0001
married*year	2	1	0.1449	0.0677	0.0122	0.2777	4.58	0.0324
married*year	3	0	0.0000	0.0000	0.0000	0.0000	.	.
happy	*1*	*1*	*0.0247*	*0.0739*	*-0.1202*	*0.1696*	*0.11*	*0.7382*
happy	*2*	*1*	*1.0899*	*0.0475*	*0.9967*	*1.1830*	*526.06*	*<.0001*
happy	*3*	*0*	*0.0000*	*0.0000*	*0.0000*	*0.0000*	*.*	*.*
happyq*year	*1*	*1*	*-0.0391*	*0.0525*	*-0.1420*	*0.0638*	*0.56*	*0.4560*

Analysis Of Maximum Likelihood Parameter Estimates								
Parameter		DF	Estimate	Standard Error	Wald 95% Confidence Limits		Wald Chi-Square	Pr > ChiSq
*happyq*year*	*2*	*1*	*-0.0900*	*0.0526*	*-0.1932*	*0.0131*	*2.93*	*0.0870*
*happyq*year*	*3*	*0*	*0.0000*	*0.0000*	*0.0000*	*0.0000*	*.*	*.*
*married*happyq*		*1*	*-0.8020*	*0.0453*	*-0.8908*	*-0.7131*	*312.97*	*<.0001*
Scale		0	1.0000	0.0000	1.0000	1.0000		

Looking first at the deviance (and Pearson chi-square), we see strong evidence that the fitted model is unsatisfactory: the deviance is more than five times the degrees of freedom. This is similar to what we found in Output 6.10 when we estimated the equivalent logit model by weighted least squares. Although there is little evidence for an effect of YEAR, the chi-square for MARRIED is quite large. The parameter estimate for the marital status effect (MARRIED*HAPPYQ) is –.8020. Again this is quite close to the WLS estimate of –.8035 in Output 6.7. Because higher values of HAPPYQ indicate lower happiness and MARRIED=1 for married, otherwise 0, the negative sign of the coefficient tells us that married people are happier. To be more specific, if we contrast the categories "very happy" and "pretty happy," the odds of a married person being in the latter category are exp(–.8020)=.45 times the odds of an unmarried person. To put it another way, married people are about twice as likely (on the odds scale) as unmarried people to be in the happier category. Because of the constraints on the model, the same statement can be made about the contrast between "pretty happy" and "not too happy"—married people are about twice as likely as unmarried people to be in the happier category.

Given the poor fit of the model, we must be wary of these interpretations. But how can the model be improved? There are two possibilities. We can allow an interaction between year and marital status in their effects on happiness, which corresponds to a three-way interaction in the loglinear model. Or, we can relax the adjacent-categories constraint and fit a regular multinomial logit model. To fit the interaction model, we use the following statement:

```
MODEL count=year|married happy year*happyq married*happyq
    married*year*happyq / D=P ;
```

That model produces a deviance of 29.64 with 5 d.f. Not much of an improvement, and still an unacceptable fit. We can fit the unrestricted multinomial model without a three-way interaction by simply deleting the Q from the HAPPY variable, thereby treating it as a CLASS variable:

```
MODEL count=year|married happy year*happy married*happy /
   D=P ;
```

This produces a deviance of 4.16 with 4 d.f., for a *p*-value of .38. (Since GENMOD does not report *p*-values for the deviance, all *p*-values in this chapter were calculated by using the PROBCHI function in a SAS DATA step.) Furthermore, there is a highly significant reduction in deviance when we move from the adjacent-categories model to the multinomial model. This suggests that either YEAR or MARRIED does not have a uniform effect on both pairs of adjacent categories. To see where the problem lies, take a look at the parameter estimates of the multinomial model in Output 10.7.

Output 10.7 Selected Parameter Estimates for the Multinomial Model

Analysis Of Maximum Likelihood Parameter Estimates									
Parameter			DF	Estimate	Standard Error	Wald 95% Confidence Limits		Wald Chi-Square	Pr > ChiSq
happy	1		1	0.0917	0.0781	-0.0614	0.2449	1.38	0.2404
happy	2		1	1.3038	0.0662	1.1740	1.4335	387.72	<.0001
happy	3		0	0.0000	0.0000	0.0000	0.0000	.	.
year*happy	1	1	1	-0.1135	0.1098	-0.3286	0.1017	1.07	0.3014
year*happy	1	2	1	-0.4025	0.1028	-0.6039	-0.2010	15.33	<.0001
year*happy	1	3	0	0.0000	0.0000	0.0000	0.0000	.	.
year*happy	2	1	1	0.0653	0.1114	-0.1530	0.2836	0.34	0.5577
year*happy	2	2	1	-0.1941	0.1028	-0.3956	0.0075	3.56	0.0591
year*happy	2	3	0	0.0000	0.0000	0.0000	0.0000	.	.
year*happy	3	1	0	0.0000	0.0000	0.0000	0.0000	.	.
year*happy	3	2	0	0.0000	0.0000	0.0000	0.0000	.	.
year*happy	3	3	0	0.0000	0.0000	0.0000	0.0000	.	.
married*happy	1		1	1.5725	0.0950	1.3863	1.7587	274.07	<.0001
married*happy	2		1	0.7092	0.0867	0.5391	0.8792	66.83	<.0001
married*happy	3		0	0.0000	0.0000	0.0000	0.0000	.	.

To save space, Output 10.7 has been edited to display only those parameter estimates that correspond to the multinomial logit model. Of course, those parameters include all model terms that contain HAPPY. In the first two coefficients for MARRIED, we see that the coefficient for category 1 (versus category 3) is 1.5725, which is about twice as large as the coefficient of .7092 for category 2. That's what we would expect from the adjacent categories model, so apparently the problem is not with this variable. However, things aren't quite so simple with YEAR. In the previous model, we found no evidence for any effect of year on happiness, but now we see that the 1974 contrast between categories 2 and 3 of HAPPY is highly significant, and the 1984 contrast between categories 2 and 3 is marginally significant. But the contrasts between categories 1 and 3 of HAPPY are both far from significant. Clearly, there is some kind of effect of year on happiness, but it doesn't conform to the adjacent categories assumption. What appears to be happening is that each 10-year increment in time is associated with an increase in the odds of being in the middle category (pretty happy) and a corresponding reduction of the odds of being in either of the two extreme categories.

We can represent this interpretation by a more parsimonious model that (a) treats happiness as quantitative in its association with marital status and (b) linearizes the effect of year on a dichotomous version of happiness—pretty happy versus the two extreme responses:

```
DATA happy3;
  SET happy2;
  yearq=year;
  pretty=happy eq 2;
PROC GENMOD DATA=happy3;
  CLASS year happy;
  MODEL count=year|married happy yearq*pretty married*happyq
    / D=P;
RUN;
```

Output 10.8 Selected Parameter Estimates for the Parsimonious Model

Analysis Of Maximum Likelihood Parameter Estimates								
Parameter		DF	Estimate	Standard Error	Wald 95% Confidence Limits		Wald Chi-Square	Pr > ChiSq
happy	1	1	0.0852	0.0645	-0.0412	0.2116	1.75	0.1865
happy	2	1	0.7303	0.0845	0.5646	0.8959	74.64	<.0001
happy	3	0	0.0000	0.0000	0.0000	0.0000	.	.
yearq*pretty		1	0.1722	0.0320	0.1095	0.2349	28.98	<.0001

Analysis Of Maximum Likelihood Parameter Estimates							
Parameter	DF	Estimate	Standard Error	Wald 95% Confidence Limits		Wald Chi-Square	Pr > ChiSq
married*happyq	1	-0.7944	0.0448	-0.8821	-0.7066	315.00	<.0001
Scale	0	1.0000	0.0000	1.0000	1.0000		

This model has a deviance of 9.66 with 8 d.f. for a *p*-value of .29, which is certainly an acceptable fit. The relevant parameter estimates are shown in Output 10.8. The coefficient for MARRIED is about what it was in the original adjacent categories model and has the same interpretation: for any contrast of adjacent categories, married people have about twice the odds of being in the happier category as unmarried people. The YEARQ*PRETTY coefficient can be interpreted as follows: with each additional decade, the odds of being in the middle category (as compared to the two extremes) rises by about 18%=100[exp(.1722)–1].

We have achieved the parsimony of this final model by carefully adjusting it to fit observed patterns that might have arisen from random variation. So, we should not feel extremely confident that this is the correct model. Nevertheless, this exercise should give you some idea of the range of possibilities for models like this. While the loglinear approach can be cumbersome, it is also remarkably flexible.

10.5 Loglinear Models for Square, Ordered Tables

Loglinear models have been particularly popular for the analysis of mobility tables like the one shown in Table 10.2, which is based on data collected in Britain by Glass (1954) and his collaborators. The column variable is the respondent's occupational status, classified into five categories with 1 being the highest and 5 the lowest. The same classification was used for the father's occupation, the row variable in the table. For obvious reasons, we call this a square table. Our aim is to fit loglinear models that describe the relationship between father's status and son's status. In the process, we'll see how to estimate some models proposed by Leo Goodman (1970), one of the pioneers of loglinear analysis.

Table 10.2 Cross-Classification of Respondent's Occupational Status by Father's Occupational Status, 3,497 British Males

		Son's Status				
		1	2	3	4	5
Father's Status	1	50	45	8	18	8
	2	28	174	84	154	55
	3	11	78	110	223	96
	4	14	150	185	714	447
	5	0	42	72	320	411

One model that nearly every researcher knows how to fit (but may not realize it) is the *independence model*, which asserts that the two variables are independent. Here is SAS code to fit the independence model:

```
DATA mobility;
  INPUT n dad son;
  DATALINES;
50   1 1
45   1 2
8    1 3
18   1 4
8    1 5
28   2 1
174  2 2
84   2 3
154  2 4
55   2 5
11   3 1
78   3 2
110  3 3
223  3 4
96   3 5
14   4 1
150  4 2
185  4 3
714  4 4
447  4 5
0    5 1
42   5 2
72   5 3
```

```
320 5 4
411 5 5
;
PROC GENMOD DATA=mobility;
  CLASS dad son;
  MODEL n = dad son /D=P;
RUN;
```

As usual, each cell of the table is read in as a separate record, with the variable N containing the frequency counts, and the variables DAD and SON containing the row and column variables with values of 1 through 5. In PROC GENMOD, these are declared as CLASS variables so that, at this point, no ordering of the categories is assumed. The MODEL statement includes the main effects of DAD and SON, but no interaction. This allows for variation in the marginal frequencies, but doesn't allow for any relationship between the two variables. Not surprisingly, the fit of this model is terrible. With 16 d.f., the deviance is 810.98, and the Pearson chi-square is 1199.36. (It's hardly worth the effort to calculate *p*-values because they would obviously be smaller than any sensible criterion.) Note that the Pearson chi-square is the same value that would be obtained by traditional methods of computing chi-square in a two-way table and the same value that is reported by PROC FREQ under the CHISQ option.

In rejecting the independence model, we conclude that there is indeed a relationship between father's status and son's status. But how can we modify the model to represent this relationship? We could fit the *saturated* model by the statement:

```
MODEL n=dad son dad*son / D=P;
```

but that wouldn't accomplish much. The deviance and Pearson chi-square would both be 0, and we'd have estimates for 16 parameters describing the relationship between the two variables. We might as well just look at the original table. Can't we get something more parsimonious?

The first alternative model that Goodman considered is the *quasi-independence model*, also called the *quasi-perfect mobility model* when applied to a mobility table. This model takes note of the fact that the main diagonal cells in Table 10.2 tend to have relatively high frequency counts. We might explain this by postulating a process of occupational inheritance such that sons take up the same occupation as the father. The quasi-independence model allows for such inheritance but asserts that there is no additional relationship between father's status and son's status. That is, if the son doesn't have the same status as the father, then father's status doesn't tell us anything about son's status.

There are two ways to fit the quasi-independence model. One way is to include a separate parameter for each of the main diagonal cells. The other, equivalent way (which we will take) is to simply delete the main diagonal cells from the data being fitted. Here's how:

```
PROC GENMOD DATA=mobility;
  WHERE dad NE son;
  CLASS dad son;
  MODEL n = dad son /D=P;
RUN;
```

In the WHERE statement, NE means "not equal to."

Although the quasi-independence model fits much better than the independence model, the fit is still bad. With 11 d.f., the deviance is 249.4 and the Pearson chi-square is 328.7. We conclude: although 69% of the original deviance is attributable to the main diagonal cells, there is something else going on in the table besides status inheritance.

To represent that something else, Goodman proposed 21 other models as possible candidates. Let's consider two of them. His *QPN* model is based on the ordering of occupational status. In Table 10.2, the upper triangle represents downward mobility, while the lower triangle represents upward mobility. The QPN model says that, besides ignoring the main diagonal cells, there is independence *within* each of these two triangles. To fit the model, we create a new variable distinguishing the two portions of the table:

```
DATA b;
  SET mobility;
  up=son GT dad;
PROC GENMOD DATA=b;
  WHERE dad NE son;
  CLASS dad son;
  MODEL n = dad son son*up dad*up / D=P;
RUN;
```

We get a considerable improvement in fit from this model (output not shown). With 6 d.f., the deviance is 14.0 (p=.03) and the Pearson chi-square is 9.9 (p=.13). While the p-value for the deviance is below the .05 criterion, keep in mind that we are working with a rather large sample and even minor deviations from the model are likely to be statistically significant. What is the substantive interpretation of this model? In addition to allowing for status inheritance (by deleting the main diagonal), it seems to be saying that the father's status could affect whether the son moves up or down but does not determine the exact destination.

Another of Goodman's models is the *diagonals parameter model*. This model is motivated by the observation that the cells in Table 10.2 that are farther away from the main diagonal tend to have smaller frequencies. To represent this, we include a distinct parameter corresponding to each absolute difference between father's status and son's status:

```
DATA c;
  SET mobility;
  band=ABS(dad-son);
PROC GENMOD DATA=c;
  WHERE dad NE son;
  CLASS dad son band;
  MODEL n = dad son band /D=P;
RUN;
```

The results are shown in Output 10.9. The fit is not terrible, but it's not great either—the *p*-value for the deviance is .014. The only parameters of interest to us are those for BAND. We see that cells that are directly adjacent to the main diagonal (BAND=1) have frequencies that are estimated to be exp(2.4824)=12 times those in the off-diagonal corners. For BAND=2 and BAND=3, the frequency counts are estimated to be 7 times and 3 times those in the corners.

Output 10.9 Results from Fitting the Diagonals Parameter Model

Criteria For Assessing Goodness Of Fit			
Criterion	DF	Value	Value/DF
Deviance	8	19.0739	2.3842
Scaled Deviance	8	19.0739	2.3842
Pearson Chi-Square	8	15.9121	1.9890
Scaled Pearson X2	8	15.9121	1.9890
Log Likelihood		8473.7383	
Full Log Likelihood		-65.6228	
AIC (smaller is better)		155.2457	
AICC (smaller is better)		199.8171	
BIC (smaller is better)		167.1944	

Analysis Of Maximum Likelihood Parameter Estimates								
Parameter		DF	Estimate	Standard Error	Wald 95% Confidence Limits		Wald Chi-Square	Pr > ChiSq
Intercept		1	3.0797	0.3680	2.3584	3.8010	70.03	<.0001
dad	1	1	-1.3915	0.1293	-1.6450	-1.1380	115.74	<.0001
dad	2	1	-0.2227	0.0802	-0.3799	-0.0655	7.71	0.0055
dad	3	1	-0.4620	0.0743	-0.6076	-0.3163	38.64	<.0001
dad	4	1	0.5450	0.0814	0.3854	0.7046	44.79	<.0001
dad	5	0	0.0000	0.0000	0.0000	0.0000	.	.
son	1	1	-2.1280	0.1478	-2.4176	-1.8384	207.39	<.0001
son	2	1	-0.5790	0.0756	-0.7273	-0.4308	58.61	<.0001
son	3	1	-0.8890	0.0726	-1.0314	-0.7467	149.81	<.0001
son	4	1	0.2367	0.0768	0.0861	0.3873	9.49	0.0021
son	5	0	0.0000	0.0000	0.0000	0.0000	.	.
band	1	1	2.4824	0.3707	1.7557	3.2090	44.83	<.0001
band	2	1	1.9539	0.3733	1.2222	2.6857	27.39	<.0001
band	3	1	1.1482	0.3740	0.4152	1.8811	9.43	0.0021
band	4	0	0.0000	0.0000	0.0000	0.0000	.	.

Before concluding our analysis, let's consider one additional model that was not discussed in Goodman's article, the *quasi-uniform association model*. This model incorporates the ordering of the occupational statuses in a very explicit manner. The conventional uniform association model says that if the cells in the table are properly ordered, then the odds ratio (cross-product ratio) in every 2×2 subtable of adjacent cells is exactly the same. Our modification of that model is to delete the main diagonals before we fit it. Here's the SAS code:

```
DATA d;
  SET mobility;
  sonq=son;
  dadq=dad;
PROC GENMOD DATA=d;
  WHERE dad NE son;
  CLASS dad son;
  MODEL n = dad son sonq*dadq /D=P;
  OUTPUT OUT=a PRED=pred LOWER=lower UPPER=upper;
PROC PRINT DATA=a;
  VAR n pred lower upper;
RUN;
```

As in some earlier examples, we define new versions of the row and column variables so that we can treat those variables as both categorical and quantitative. The model includes the main effects of SON and DAD as categorical, thereby ensuring that the marginal frequencies are fitted exactly. The relationship between the two variables is specified by an interaction between the quantitative versions of the two variables. The OUTPUT statement requests predicted values with confidence intervals. PROC PRINT writes them to the output window, as shown in Output 10.10.

With 10 d.f., the quasi-uniform association model has a deviance of 19.3 (p=.04) and a Pearson chi-square of 16.2 (p=.094) (output not shown). While not quite as good a fit as the QPN model, it's still decent for a sample of this size, and it has the virtue of providing a single number to describe the association between father's status and son's status in the off-diagonal cells: .3374. Exponentiating, we find that the estimated odds ratio in any 2×2 subtable is 1.40. Thus for adjacent categories, being in the higher category for DAD increases the odds of being in the higher category for SON by 40%.

Comparing the observed and predicted frequencies in Output 10.10, we find that the numbers are close for most of the cells. (For the three lines shown in italics, the observed frequency lies outside the 95% confidence interval based on the fitted model.)

Output 10.10 Selected Output for the Quasi-Uniform Association Model

Obs	n	pred	lower	upper
1	45	36.295	28.089	46.898
2	*8*	*16.126*	*12.634*	*20.585*
3	18	21.378	16.327	27.990
4	*8*	*5.201*	*3.696*	*7.319*
5	28	24.794	18.394	33.421
6	84	84.848	72.629	99.123
7	154	157.617	138.653	179.175
8	55	53.741	45.198	63.899
9	11	11.808	8.866	15.726
10	78	90.944	78.320	105.603
11	223	206.557	184.038	231.832
12	96	98.691	85.919	113.362
13	14	13.911	10.139	19.087
14	150	150.146	131.588	171.321
15	185	183.577	162.598	207.262
16	447	448.366	410.689	489.499

Obs	n	pred	lower	upper
17	*0*	*2.487*	*1.688*	*3.664*
18	42	37.616	31.135	45.445
19	72	64.449	55.153	75.310
20	320	329.449	297.831	364.422

10.6 Marginal Tables

To understand the literature on loglinear models, it's helpful to have some appreciation of *marginal tables* and their relationship to the parameters in the loglinear model and to the maximum likelihood estimates. A marginal table is just a contingency table that is obtained by summing the frequencies in some larger contingency table over one or more of the variables. For example, if we have a three-way table for variables *X*, *Y*, and *Z*, there are several possible marginal tables:

- the $X \times Y$ table, obtained by summing over *Z*
- the $X \times Z$ table, obtained by summing over *Y*
- the $Y \times Z$ table, obtained by summing over *X*
- the *X* table, obtained by summing over both *Y* and *Z*
- the *Y* table, obtained by summing over both *X* and *Z*
- the *Z* table, obtained by summing over both *Y* and *X*

Why are these tables important? Because every term in a loglinear model has a corresponding marginal table. For example, to fit a saturated model for the $X \times Y \times Z$ table in GENMOD, we specify:

```
MODEL f = x y z x*y x*z y*z x*y*z / D=P;
```

Each term in this model corresponds to a specific table or subtable, with the three-way interaction corresponding to the full three-way table. Because this is a saturated model, the predicted cell frequencies are identical to the observed cell frequencies in the full table. In the notation of Section 10.2 (with *i* representing a value of variable *X*, *j* representing a value of *Y*, and *k* representing a value of *Z*), we have $\hat{m}_{ijk} = n_{ijk}$ for all *i*, *j*, and *k*. Clearly, this must also be true for any marginal table: the summed predicted frequencies will be equal to the summed observed frequencies.

Now consider an unsaturated model that deletes the three-way interaction:

```
MODEL f = x y z x*y x*z y*z / D=P;
```

For this model, the maximum likelihood estimates of the expected frequencies in the three-way table will not, in general, be equal to the observed frequencies: $\hat{m}_{ijk} \neq n_{ijk}$. However, for all the marginal tables corresponding to the terms in the model, the summed predicted frequencies will equal the summed observed frequencies. (The + signs in the following subscripts indicate that the frequencies are summed over that variable.)

$$\begin{aligned}
\hat{m}_{ij+} &= n_{ij+} \\
\hat{m}_{i+k} &= n_{i+k} \\
\hat{m}_{+jk} &= n_{+jk} \qquad \text{for all } i, j, \text{ and } k \\
\hat{m}_{i++} &= n_{i++} \\
\hat{m}_{+j+} &= n_{+j+} \\
\hat{m}_{++k} &= n_{++k}
\end{aligned}$$

You can readily see this with the independence model for a 2 × 2 table. Table 10.3 displays observed and predicted frequencies for the independence model applied to Table 10.1. Clearly, both the observed and predicted frequencies sum to the one-way marginal totals.

Table 10.3 Observed (and Predicted) Frequencies for the Independence Model

	Blacks	Nonblacks	Total
Death	28 (24.8)	22 (25.2)	50
Life	45 (48.2)	52 (48.8)	97
Total	73	74	147

One final property of marginal tables is this: the marginal tables that correspond to the terms in the loglinear model are the *sufficient statistics* for that model. That means that the maximum likelihood estimates can be calculated using only the information in the appropriate marginal tables. For example, in Table 10.3 the maximum likelihood estimate of the expected frequency in each cell is obtained by multiplying the row total times the column total and dividing by the grand total (147). The parameter estimates can then be obtained from the estimated expected frequencies. So we only need the marginal tables to fit the independence model. Of course, to calculate a goodness-of-fit chi-square, we need the full table so that we can compare the expected frequencies with the observed frequencies.

For all these reasons, the estimation of a loglinear model is often referred to as a process of *fitting marginal tables*. Many journal articles that report results of a loglinear analysis describe the fitted models by specifying the fitted marginal tables for each model, not by specifying the parameters included in the model.

10.7 The Problem of Zeros

Contingency tables sometimes have cells with frequency counts of zero. These may cause problems or require special treatment. There are two kinds of zeros:

- *Structural zeros*: These are cells for which a nonzero count is impossible because of the nature of the phenomenon or the design of the study. The classic example is a cross-tabulation of sex by type of surgery in which structural zeros occur for male hysterectomies and female vasectomies.
- *Random zeros*: In these cells, nonzero counts are possible (at least as far we know), but a zero occurs because of random variation. Random zeros are especially likely to arise when the sample is small and the contingency table has many cells.

Structural zeros are easily accommodated with PROC GENMOD. Simply delete the structural zeros from the data set before estimating the model. Random zeros can be a little trickier. Most of the time they don't cause any difficulty, except that the expected cell counts may be very small, thereby degrading the chi-square approximation to the deviance and Pearson's statistic. However, more serious problems arise when random zeros show up in fitted marginal tables. *When a fitted marginal table contains a frequency count of zero, at least one ML parameter estimate is infinite and the fitting algorithm will not converge.* We already encountered this problem in Section 3.4 for the binary logit model. There, we saw that if there is a 0 in the 2×2 table describing the relationship between the dependent variable and any dichotomous independent variable, the coefficient for that variable is infinite (more accurately, the ML estimate does not exist) and the algorithm will not converge. The identical problem arises when fitting a logit model by means of its equivalent loglinear model, and the potential solutions are the same. However, problems arise more frequently when fitting loglinear models because it's necessary to fit the full marginal table describing the relationships among all the independent variables. As we've seen, loglinear models typically contain many nuisance parameters, any of which could have infinite estimates causing problems with convergence.

Here's a simple, hypothetical example. Consider the following three-way table for dichotomous variables *X*, *Y*, and *Z*:

	X = 1		*X* = 0	
	Z		*Z*	
Y	1	0	1	0
1	20	5	4	0
0	5	5	11	0
Total	25	10	15	0

Considering the two-way marginal tables, neither the *XY* table nor the *YZ* table has any zeros. But the *XZ* table clearly has one random zero, producing two random zeros in the three-way table.

Now suppose we want to estimate a logit model for *Y* dependent on *X* and *Z*. We can read the table into SAS as follows:

```
DATA zero;
  INPUT x y z f;
  DATALINES;
1    1    1    20
1    0    1     5
1    1    0     5
1    0    0     5
0    1    1     4
0    0    1    11
0    1    0     0
0    0    0     0
;
```

We can estimate the logit model directly with

```
PROC GENMOD DATA=zero DESC;
  FREQ f;
  MODEL y = x z / D=B AGGREGATE;
RUN;
```

The AGGREGATE option requests the deviance and Pearson chi-squares. This program produces the results in Output 10.11. There is no apparent problem here. The coefficient for *X* is large and statistically significant, while the coefficient for *Z* is smaller and not quite significant. Both goodness-of-fit statistics are 0 with 0 degrees of freedom. That's because the model has three parameters, but there are only three combinations of *X* and *Z* for which we observe the dependent variable *Y*.

Output 10.11 Logit Output for Data with Marginal Zeros

Criteria For Assessing Goodness Of Fit			
Criterion	DF	Value	Value/DF
Deviance	0	0.0000	.
Scaled Deviance	0	0.0000	.
Pearson Chi-Square	.	0.0000	.
Scaled Pearson X2	.	0.0000	.
Log Likelihood		-28.1403	
Full Log Likelihood		-4.5114	
AIC (smaller is better)		15.0228	
AICC (smaller is better)		15.5446	
BIC (smaller is better)		20.7589	

Analysis Of Maximum Likelihood Parameter Estimates							
Parameter	DF	Estimate	Standard Error	Wald 95% Confidence Limits		Wald Chi-Square	Pr > ChiSq
Intercept	1	-2.3979	0.9954	-4.3489	-0.4469	5.80	0.0160
x	1	2.3979	0.7687	0.8913	3.9045	9.73	0.0018
z	1	1.3863	0.8062	-0.1939	2.9665	2.96	0.0855

Now, let's estimate the equivalent loglinear model:

```
PROC GENMOD DATA=zero;
  MODEL f=x z y x*z y*z y*x / D=P AGGREGATE;
  OUTPUT OUT=a PRED=pred;
RUN;
```

The OUTPUT statement produces predicted values, which will be used below. As Output 10.12 shows, all the parameters pertaining to *Y* (and associated statistics) are the same in both the loglinear and logit versions of the model. But, all the nuisance parameters are very large, with huge chi-squares. The goodness of fit chi-squares are again 0, but the reported degrees of freedom is 1 rather than 0.

Output 10.12 Loglinear Output for Data with Marginal Zeros

Criteria For Assessing Goodness Of Fit			
Criterion	DF	Value	Value/DF
Deviance	1	0.0000	0.0000
Scaled Deviance	1	0.0000	0.0000
Pearson Chi-Square	1	0.0000	0.0000
Scaled Pearson X2	1	0.0000	0.0000
Log Likelihood		65.9782	
Full Log Likelihood		-11.4002	
AIC (smaller is better)		36.8004	
AICC (smaller is better)		.	
BIC (smaller is better)		37.3565	

Analysis Of Maximum Likelihood Parameter Estimates							
Parameter	DF	Estimate	Standard Error	Wald 95% Confidence Limits		Wald Chi-Square	Pr > ChiSq
Intercept	1	-23.6743	0.7006	-25.0476	-22.3011	1141.71	<.0001
x	1	25.2838	0.5394	24.2267	26.3409	2197.49	<.0001
z	1	26.0722	0.6325	24.8327	27.3118	1699.40	<.0001
y	1	-2.3979	0.9954	-4.3489	-0.4469	5.80	0.0160
x*z	0	-26.0722	0.0000	-26.0722	-26.0722	.	.
z*y	1	1.3863	0.8062	-0.1939	2.9665	2.96	0.0855
x*y	1	2.3979	0.7687	0.8913	3.9045	9.73	0.0018
Scale	0	1.0000	0.0000	1.0000	1.0000		

The four parameter estimates greater than 20 all stem from the 0 in the marginal table for *X* and *Z*. While there's no guarantee, my experience with PROC GENMOD is that it invariably produces the right estimates and standard errors for the parameters that do *not* pertain to the marginal table with zeros. However, other software may not do the same. Even if the logit parameter estimates are correct, the incorrect degrees of freedom for the deviance and Pearson statistics may invalidate comparisons with other models.

The solution is to treat the random zeros that arise from marginal zeros as if they were structural zeros—that is, delete them from the data before fitting the model. How do we know which random zeros in the full table come from zeros in the fitted marginal tables? In this example, it's fairly evident, but more complicated tables may present some difficulties. One approach is to compare observed with expected frequencies and check for cells in which the observed frequency is 0 and the estimated frequency is either 0 or extremely small.

Output 10.13 shows the observed and predicted values produced by the model in Output 10.12. We see that the last two lines do, in fact, have observed frequencies of 0 and predicted frequencies of 0.

Output 10.13 Observed and Predicted Values for Model with Marginal Zeros

Obs	x	y	z	f	pred
1	1	1	1	20	20
2	1	0	1	5	5
3	1	1	0	5	5
4	1	0	0	5	5
5	0	1	1	4	4
6	0	0	1	11	11
7	0	1	0	0	0
8	0	0	0	0	0

Now let's refit the model without these two zeros:

```
PROC GENMOD DATA=zero;
  WHERE f NE 0;
  MODEL f=x z y x*z y*z y*x / D=P AGGREGATE;
RUN;
```

Results in Output 10.14 give the correct logit parameter estimates and the correct degrees of freedom, 0. None of the estimated parameters is unusually large. Note, however, that we do not get an estimate for the X*Z interaction, because we've eliminated one component of the *XZ* table. Deletion of cells with zero frequency should only be used when the parameters corresponding to the marginal tables with zeros are nuisance parameters. Otherwise, follow the strategies discussed in Section 3.4.

Output 10.14 Loglinear Output for Model with Zeros Deleted

Criteria For Assessing Goodness Of Fit			
Criterion	DF	Value	Value/DF
Deviance	0	0.0000	.
Scaled Deviance	0	0.0000	.
Pearson Chi-Square	.	0.0000	.
Scaled Pearson X2	.	0.0000	.
Log Likelihood		65.9782	
Full Log Likelihood		-11.4002	
AIC (smaller is better)		34.8004	

Criteria For Assessing Goodness Of Fit			
Criterion	DF	Value	Value/DF
AICC (smaller is better)		.	
BIC (smaller is better)		33.5510	

Analysis Of Maximum Likelihood Parameter Estimates							
Parameter	DF	Estimate	Standard Error	Wald 95% Confidence Limits		Wald Chi-Square	Pr > ChiSq
Intercept	1	2.3979	0.7006	1.0246	3.7711	11.71	0.0006
x	1	-0.7885	0.5394	-1.8456	0.2687	2.14	0.1438
z	1	-0.0000	0.6325	-1.2396	1.2396	0.00	1.0000
y	1	-2.3979	0.9954	-4.3489	-0.4469	5.80	0.0160
x*z	0	0.0000	0.0000	0.0000	0.0000	.	.
z*y	1	1.3863	0.8062	-0.1939	2.9665	2.96	0.0855
x*y	1	2.3979	0.7687	0.8913	3.9045	9.73	0.0018

10.8 GENMOD versus CATMOD

I've used PROC GENMOD exclusively in this chapter, but you can also estimate loglinear models with PROC CATMOD. I prefer GENMOD for several reasons:

- PROC GENMOD can fit a wider range of loglinear models than PROC CATMOD. For example, PROC CATMOD does not allow the device of treating a variable as both quantitative and qualitative in the same loglinear model.
- PROC GENMOD can correct for overdispersion when appropriate.
- PROC GENMOD can optionally produce likelihood-ratio hypothesis tests, while PROC CATMOD only produces Wald tests.

There is one other noteworthy difference: PROC CATMOD can take individual-level data as input. It then constructs the necessary contingency table required for each model it estimates. PROC GENMOD requires the contingency table as input, although this is easily produced with PROC FREQ by using the OUT= option in the TABLES statement.

References

Agresti, A. (2002), *Categorical Data Analysis*, Second Edition. New York: John Wiley & Sons.

Allison, P. D. (1982), "Discrete-Time Methods for the Analysis of Event Histories," in *Sociological Methodology 1982*, ed. S. Leinhardt. San Francisco, CA: Jossey-Bass, 61-98.

Allison, P. D. (1987), "Introducing a Disturbance into Logit and Probit Regression Models," *Sociological Methods and Research*, 15, 355-374.

Allison, P. D. (1999), "Comparing Logit and Probit Coefficients Across Groups." Forthcoming in *Sociological Methods and Research*.

Allison, P. D. (2005), *Fixed Effects Regression Methods for Longitudinal Data Using SAS.* Cary, NC: SAS Institute Inc.

Allison, P. D. (2010), *Survival Analysis Using SAS: A Practical Guide,* Second Edition. Cary, NC: SAS Institute Inc.

Allison, P. D. and Christakis, N. A. (1994), "Logit Models for Sets of Ranked Items," in *Sociological Methodology 1994*, ed. P. V. Marsden. Oxford: Basil Blackwell, 199-228.

Begg, C. B. and Gray, R. (1984), "Calculation of Polychotomous Logistic Regression Parameters Using Individualized Regressions," *Biometrika*, 71, 11-18.

Bhat C. R. (1995) "A Heteroscedastic Extreme Value Model of Intercity Travel Mode Choice," *Transportation Research Part B: Methodological*, 29, 471-483.

Breslow, N. and Day, N. E. (1980), *Statistical Methods in Cancer Research. Vol. 1: The Analysis of Case-Control Studies.* Lyon: IARC.

Bryk, A. S. and Raudenbusch, S. W. (1992), *Hierarchical Linear Models: Applications and Data Analysis*. Newbury Park, CA: Sage.

Chamberlain, G. (1980), "Analysis of Covariance with Qualitative Data," *Review of Economic Statistics*, 48, 225-238.

Christakis, N. A. and Levinson, W. (1998), "Casual Optimism: Prognostication in Routine Medical and Surgical Encounters." Unpublished manuscript.

Collett, D. (1991), *Modelling Binary Data*. London: Chapman & Hall.

Cox, D. R. (1970), *Analysis of Binary Data*. London: Methuen.

Cox, D. R. (1972), "Regression Models and Life Tables" (with discussion), *Journal of the Royal Statistical Society, Series B*, 34, 187-220.

Cox, D. R. and Snell, E. J. (1989), *Analysis of Binary Data*, Second Edition. London: Chapman & Hall.

Davis, C. E.; Hyde, J. E.; Bangdiwala, S. I.; and Nelson, J. J. (1986), "An Example of Dependencies Among Variables in a Conditional Logistic Regression," in *Modern Statistical Methods in Chronic Disease Epidemiology*, eds. S. H. Moolgavkar and Ross L. Prentice. New York: John Wiley & Sons.

Diggle, P. J.; Liang, K. Y.; and Zeger, S. L. (1994), *The Analysis of Longitudinal Data*. New York: Oxford University Press.

Everitt, B. S. (1992), *The Analysis of Contingency Tables*, Second Edition. London: Chapman & Hall.

Fienberg, S. E. (2007), *The Analysis of Cross-Classified Categorical Data*, Second Edition. Cambridge, MA: The MIT Press.

Firth, D. (1993), "Bias Reduction of Maximum Likelihood Estimates," *Biometrika*, 80, 27-38.

Fox, John (1991), *Regression Diagnostics.* Newbury Park, CA: Sage Publications.

Gail, M. H.; Wieand, S.; and Piantadosi, S. (1984), "Biased Estimates of Treatment Effect in Randomized Experiments with Nonlinear Regression and Omitted Covariates," *Biometrika*, 71, 431-444.

Glass, D. V., ed. (1954), *Social Mobility in Britain.* Glencoe, IL: Free Press.

Goodman, L. A. (1970), "Some Multiplicative Models for the Analysis of Cross Classified Data," *Proceedings of the Sixth Berkeley Symposium on Mathematical Statistics and Probability*, 649-696.

Greene, W. H. (1992), *LIMDEP User's Manual and Reference Guide, Version 6.0.* Bellport, NY: Econometric Software, Inc.

Hauck, W. W. and Donner, A. (1977), "Wald's Test as Applied to Hypotheses in Logit Analysis," *Journal of the American Statistical Association*, 72, 851-853.

Heckman, J. J. (1978), "Dummy Endogenous Variables in a Simultaneous Equation System," *Econometrica*, 46, 931-960.

Heckman, J. J. (1979), "Sample Selection Bias as a Specification Error," *Econometrica*, 47, 153-161.

Heinze, G. and Schemper, M. (2002), "A Solution to the Problem of Separation in Logistic Regression," *Statistics in Medicine*, 21, 2409–2419.

Hensher, D. A. and Bradley, M. (1993), "Using Stated Response Data to Enrich Revealed Preference Discrete Choice Models," *Marketing Letters*, 4, 139-152.

Hilbe, J. (1994), "Log Negative Binomial Regression Using the GENMOD Procedure SAS/STAT Software," Proceedings of SUGI 19. Cary, NC: SAS Institute Inc.

Hin, L .Y. and Wang, Y. G. (2009), "Working-Correlation-Structure Identification in Generalized Estimating Equations," *Statistics in Medicine*, 28, 642–658.

Horney, J.; Osgood, D. W.; and Marshall, I. H. (1995), "Criminal Careers in the Short-Term: Intra-Individual Variability in Crime and Its Relation to Local Life Circumstances," *American Sociological Review*, 60, 655-673.

Hosmer, D. W. and Lemeshow, S. (2000), *Applied Logistic Regression*, 2nd Edition. New York: John Wiley & Sons.

Hout, M. (1983), *Mobility Tables*. Beverly Hills: Sage Publications.

Huber, P. J. (1967) "The Behavior of Maximum Likelihood Estimates Under Nonstandard Conditions," in *Proceedings of the Fifth Berkeley Symposium in Mathematical Statistics*, Volume 1. Berkeley: University of California Press, 221-233.

Jennings, D. E. (1986), "Judging Inference Adequacy in Logistic Regression," *Journal of the American Statistical Association*, 81, 471-476.

Kalbfleisch, J. D. and Sprott, D. A. (1970), "Applications of Likelihood Methods to Models Involving Large Numbers of Parameters" (with discussion), *Journal of the Royal Statistical Society, Series B*, 32, 175-208.

Kauermann, G. and Carroll, R. J. (2001), "A Note on the Efficiency of Sandwich Covariance Matrix Estimation," *Journal of the American Statistical Association*, 96: 1387-1396.

Keane, A.; Jepson, C.; Pickett, M.; Robinson, L; and McCorkle, R. (1996), "Demographic Characteristics, Fire Experiences and Distress of Residential Fire Survivors," *Issues in Mental Health Nursing*, 17, 487-501.

Kim, J. and Feree, G. D. (1981), "Standardization in Causal Analysis," *Sociological Methods and Research*," 10, 187-210.

Levinson, W.; Roter, D. L.; Mullooly, J. P.; Dull, V. T.; and Frankel, R. M. (1997), "Physician-Patient Communication," *Journal of the American Medical Association*, 277, 553-559.

Long, J. S. (1997), *Regression Models for Categorical and Limited Dependent Variables*. Thousand Oaks, CA: Sage Publications.

Long, J. S.; Allison, P.D.; and McGinnis, R. (1993) "Rank Advancement in Academic Careers: Sex Differences and the Effects of Productivity," *American Sociological Review* 58, 703-722.

Maddala, G. S. (1983), *Limited Dependent and Qualitative Variables in Econometrics*. Cambridge, UK: Cambridge University Press.

McCullagh, P. (1980), "Regression Models for Ordinal Data" (with discussion), *Journal of the Royal Statistical Society, Series B*, 42, 109-142.

McCullagh, P. and Nelder, J. A. (1989), *Generalized Linear Models*, Second Edition. London: Chapman and Hall.

McCulloch, C. E. (1997), "Maximum Likelihood Algorithms for Generalized Linear Mixed Models," *Journal of the American Statistical Association*, 92, 162-170.

McFadden, D. (1974), "Conditional Logit Analysis of Qualitative Choice Behavior," in *Frontiers in Econometrics*, ed. by P. Zarembka. New York: Academic Press, 105-142.

McGinnis, R.; Allison, P. D.; and Long, J. S. (1982), "Postdoctoral Training in Bioscience: Allocation and Outcomes," *Social Forces*, 60, 701-722.

Metraux, S. and Culhane, D. P. (1999) "Family Dynamics, Housing and Recurring Homelessness Among Women in New York City Homeless Shelters," *Journal of Family Issues* 20.3, 371-398.

Morgan, S. P. and Teachman, J. D. (1988), "Logistic Regression: Description, Examples and Comparisons," *Journal of Marriage and the Family*, 50, 929-936.

Muthén, B. (1984), "A General Structural Equation Model with Dichotomous, Ordered, Categorical, and Continuous Latent Variable Indicators," *Psychometrika*, 49, 115-132.

Nagelkerke, N. J. D. (1991), "A Note on a General Definition of the Coefficient of Determination," *Biometrika*, 78, 691-692.

Neuhaus, J. M. and Kalbfleisch, J. D. (1998), "Between- and Within-Cluster Covariate Effects in the Analysis of Clustered Data," *Biometrics*, 54, 638-645.

Pan, W. (2001), "Akaike's Information Criterion in Generalized Estimating Equations," *Biometrics*, 57, 120-125.

Prentice, R. and Pyke, R. (1979), "Logistic Disease Incidence Models and Case-Control Studies," *Biometrika*, 66, 403-412.

Punj, G. N. and Staelin, R. (1978), "The Choice Process for Graduate Business Schools," *Journal of Marketing Research*, 15, 588-598.

Rodriguez, G. and Goldman, N. (1995), "An Assessment of Estimation Procedures for Multilevel Models with Binary Responses," *Journal of the Royal Statistical Society, Series A*, 158, 73-89.

Rosenbaum, P. R. and Rubin, D. B. (1983), "The Central Role of the Propensity Score in Observational Studies for Causal Effects," *Biometrika*, 70, 41-55.

SAS Institute Inc. (1995), *Logistic Regression Examples Using the SAS System.* Cary, NC: SAS Institute Inc.

Seeman, M. (1977), "Some Real and Imaginary Consequences of Social Mobility: A French-American Comparison," *American Journal of Sociology*, 82, 757-782.

Sewell, W. H. and Shah, V. P. (1968), "Parents' Education and Children's Educational Aspirations and Achievements," *American Sociological Review*, 33, 191-209.

Silberhorn, N.; Boztug, Y.; and Hildebrandt, L. (2008), "Estimation with the Nested Logit Model: Specifications and Software Particularities," *OR Spectrum*, 30:635–653.

Sloane, D. and Morgan, S. P. (1996), “An Introduction to Categorical Data Analysis,” *Annual Review of Sociology*, 22, 351-375.

Smith, H. L. (1997), “Matching With Multiple Controls to Estimate Treatment Differences in Observational Studies,” in *Sociological Methodology 1997*, ed. A. E. Raftery. Oxford: Basil Blackwell, 325-353.

Stokes, M. E.; Davis, C. S.; and Koch, G. (2000), *Categorical Data Analysis Using the SAS System,* Second Edition. Cary, NC: SAS Institute Inc.

Tjur, T. (2009), “Coefficients of Determination in Logistic Regression Models—A New Proposal: The Coefficient of Discrimination,” *The American Statistician*, 63, 366-372.

Veall, M. R. and Zimmermann, K. F. (1996), “Pseudo-R2 Measures for Some Common Limited Dependent Variable Models,” *Journal of Economic Surveys*, 10, 241-259.

White, H. A. (1980), “A Heteroskedasticity-Consistent Covariance Matrix Estimator and a Direct Test for Heteroskedasticity,” *Econometrica*, 48, 817-838.

Williams, D. A. (1982), “Extra-binomial Variation in Logistic Linear Models,” *Applied Statistics*, 31, 144–148.

Index

D

E

F

G

H

I

L

M

N

O

P

Made in the USA
Lexington, KY
20 July 2013